Wir danken dem Sponsor

Wir danken dem Sponsor

Glaxo
PULMOLOGICA

Hartmut Zwick (ed.)

Sleep Related Breathing Disorders

1992. Approx. 100 pages.
Soft cover DM 38,–, öS 266,–
ISBN 3-211-82376-X

Prices are subject to change without notice

Sleep related breathing disorders have been known for many years to be an important cause of daytime disability. This book supplies an overview of diagnostic and therapeutic approaches for this disease. A lecture on sleep in normal subjects, patients with obstructive pulmonary diseases with special emphasis on the combination of COPD with coexisting sleep apnea by J.N. Douglas is followed by an article about pulmonary hemodynamics in sleep apnea by J. Kriegler. J.H. Peters gives a lecture upon epidemiology of sleep related breathing disorders, H. Rauscher presents a stepwise diagnostic procedure to sleep apnea. J.R. Stradling makes some fairly controversal statements about current treatment and comes on to describe the nasal CPAP. The book is completed by Round table discussion about current and future perspectives in sleep apnea.

Springer-Verlag Wien New York

Friedrich Kummer (Hrsg.)

Das cholinerge System der Atemwege

1992. 56 Abb. (2 davon in Farbe).
VII, 165 Seiten.
Broschiert öS 345,-, DM 49,-
ISBN 3-211-82341-7

Preisänderungen vorbehalten

Das cholinerge System stellt eines der Hauptregulative des bronchialen Muskeltonus dar. Es spielt bei verschiedenen obstruktiven Ventilationsstörungen eine jeweils andere – wichtige oder untergeordnete – Rolle.

Es wird in diesem Buch die Brücke zwischen der Theorie (neuronale Strukturen, Rezeptoren, Mediatoren) und der klinischen Praxis geschlagen. Die Besonderheit des Buches liegt u. a. darin, daß erstmals die traditionsreiche europäische Forschung auf diesem Gebiet (W. T. Ulmer, Bochum) mit den neueren sehr klinisch orientierten Ergebnissen der Autoren aus USA (N. Gross) und Canada (Chapmann) gemeinsam präsentiert wird. Dies regt den vorgebildeten Leser zur Abwägung und zum Vergleich an, während Laien auf dem Gebiet der Atemphysiologie gleich in das Zentrum des Interesses eingeführt werden.

Das Buch vermittelt einen hochmodernen Überblick über unser Verständnis des Vagus bzw. der cholinerg-mediierten Bronchokonstriktion jeweils mit konkretem klinischen Bezug auf allergisches Asthma, kindlichem Bronchospasmus, chronische Bronchitis und Emphysem. Den therapeutischen Konsequenzen wird breiter Raum gegeben. Ein spezielles Interesse wird der Interaktion von Linksherzinsuffizienz und bronchialer Hyperreaktivität (A. Lockhart, Paris) bzw. bronchialer Obstruktion gewidmet, ein wieder sehr zukunftsträchtiges Forschungsgebiet.

Springer-Verlag Wien New York

Hartmut Zwick (ed.)

Bronchial Hyperresponsiveness

Springer-Verlag Wien New York

Primarius Dr. Hartmut Zwick

Vorstand der Lungenabteilung
Krankenhaus der Stadt Wien-Lainz, Wien, Austria

Wiener Straße 80, 3580 Horn
Printed on acid-free paper

With 17 Figures

ISBN-13:978-3-211-82375-0   e-ISBN-13:978-3-7091-9228-3
DOI: 10.1007/978-3-7091-9228-3

Foreword

The Austrian Pneumological Society held its 33th Workshop on "Clinical Respiratory Physiology" at Graz, November 1st - 3rd, 1990.

Bronchial hyperresponsiveness has been known nearly as long as asthma bronchiale itself. During the last two decades while exploring the inflammatory nature of asthma bronchiale we have learned a lot about measuring and modifying this phenomenon. Because of the practical relevance and owing to the high competence of the authors we hope that the lectures and discussions we had at Graz will inform and enjoy the interested reader.

I want to express my deep gratitude to all the authors for providing us with the manuscripts and graphs. Finally I am especially grateful to Mrs. H. Weber for her secreterial work and to Mrs. Mag. A. Lahrmann-Ramharter for correcting the written version of the speeches and preparing a manuscript that was ready for the press.

Prim. Dr. Hartmut Zwick

Contents

Cellular and Biochemical Aspects of Airway Hyperresponsiveness

Peter Barnes

National Heart & Lung Institute, London (UK)

I think in the last two years the concepts of asthma have changed rather strikingly and most people would now accept asthma as an inflammatory condition.

If you look down a bronchoscope of a normal person you can see that the airway mucosa is rather pale and sharply defined. By contrast asthmatic airways are reddened and swollen. These are the classic signs of inflammation. If you biopsy the walls of asthmatic airways you can see evidence of an inflammatory reaction, infiltration with inflammatory cells and edematous swelling of the airway. An important question now is: what are the cellular components of this inflammatory reaction, and how do the products of inflammatory cells lead to the characteristic features of asthma, such as airway hyperresponsiveness and the symptoms of chronic asthma?

In the past, I think, people assumed that asthma would be explained by some abnormality of airway smooth muscle, since the symptoms of asthma come on very quickly and can be reversed rapidly by bronchodilators which act mainly by relaxing airway smooth muscle. So much research initially concentrated on the possibility that airway smooth muscle was abnormal in asthma. But despite extensive research, I think there is little convincing evidence to support this possibility. If asthmatic airways are studied in a muscle bath, the contractile response of the muscle to spasmogens like histamine and acetylcholine is quite normal. There is neither an increased sensitivity to spasmogens nor an increased amount of contraction. It therefore seems unlikely that there is a fundamental abnormality in airway smooth muscle itself, and more likely that there is an abnormality in the control of airway caliber making the airways narrow more extensively to stimuli which would not normally be able to affect airway caliber. Research is now concentrating on the components of this inflammatory reaction which in some way must make asthmatic airways narrow too much and too readily.

It now seems that inflammation of the airway can lead to the Increased airway responsiveness which is so characteristic of asthma. Once the airways become hyperresponsive, this allows triggers which would normally be ineffective to produce symptoms of asthma through airway narrowing. However, the relationship between inflammation, reactivity and symptoms, is not as clear as was previously believed. Treatments which are anti-inflammatory, particularly inhaled steroids, which suppress the inflammation, can dramatically relieve the symptoms of asthma while having a relatively small effect on airway reactivity. This suggests that inflammation itself can directly lead to the symptoms of asthma. We believe that this is due to sensitization and activation of sensory nerve endings in the airway, and, of course, pain is a classical sign of inflammation. The equivalents of pain in asthma may be symptoms such as cough and the feeling of tightness,   which are such characteristic symptoms of chronic asthma.

One of the models which has been used extensively in trying to understand the inflammation of asthma has been airway challenge. This has told us a lot of asthma, but now perhaps is being overinterpreted. As you know, allergen challenge is performed by exposing asthmatic patients to a high concentration of nebulized allergen. The concentration is probably higher than under any natural conditions. This leads to an immediate fall in lung function which can be readily prevented and reversed by bronchodilators, but is frequently followed by a later fall in lung function which tends to come on several hours after the initial allergen exposure. This "late response" is much less readily reversed by bronchodilators, and therefore the narrowing of the airways is due to more than spasm of airway smooth muscle. Perhaps the most likely explanation is that it is due to swelling of the airway in addition to bronchospasm. It is interesting that during this late response the patient may feel ill, and may feel that airway function is worse than during the early response, even though the spirometric measurements would indicate that the airway narrowing is less severe. This may be because during this late response there is an acute inflammatory reaction in the airways with the release of cytokines. The cytokines that are released cause systemic symptoms such as fever. Following this late response the airways may behave abnormally for several days and during this time the patient may have increased symptoms of asthma, such as a fall in lung function at night, nocturnal asthma, exercise induced asthma, coughing. This appears to be linked to a period of an increased responsiveness. So exposure to allergen on a single occasion can lead to very long lasting events in the airways and can exacerbate asthma. So it is important to try to understand the mechanisms for this longterm exacerbation of asthma, since they are more likely to be relevant to chronic asthma than the early events which occur immediately after allergen challenge.

The explanation that was put forward to explain asthma in the past was that mast cells are activated by exposure to allergen through high affinity IgE receptors, which means that very small concentrations of allergen are necessary or are capable of triggering these cells to release mediators such as histamine. Histamine acts on receptors on airway smooth muscle leading to bronchial constriotion, and so asthma could be treated by giving bronchodilators to prevent this bronchial constriction. But since mediators have effects in addition to contracting airway smooth muscle, a more fundamental approach to the treatment of asthma would be to find drugs that stabilize mast cells to prevent them from releasing mediators. So attractive was this possibility that many major pharmaceutical companies searched for mast cell stabilizing drugs. Over 40 different compounds were found, many of which were highly effective as mast cell stabilizers, but none has been found to work in patients with asthma, which is a powerful argument against the central role of the mast cells.

Even more powerful evidence against the central role of mast cells is provided by the effects of the drugs which are now most effective in the treatment of asthma. Beta-agonists are effective   mast cell stabilizing drugs, and block the early response to allergens. Yet when given before the allergen   challenges are not very effective in blocking the late response or the subsequent bronchial hyperresponsiveness. Indeed in patients with asthma it is not possible to significantly reduce the airway hyperresponsiveness by giving beta-agonists over a long time. By contrast corticosteroids have no effect on the early response, which suggests that human mast cells have no steroid receptors. Inhaled steroids even in a single dose can block the late response to allergen, the subsequent bronchial hyperresponsiveness, and can reverse or at least reduce the airway hyperresponsiveness of asthma. So we can conclude that mast cells are likely to be involved in the immediate response to allergen and probably to other indirect bronchial constrictors like exercise and fog, but they cannot play a critical role in either the late response or airway hyperresponsiveness, which we would now regard as more relevant to understanding chronic day-to-day asthma.

For that reason people have looked at other inflammatory cells which might be more important in these chronic inflammatory mechanisms. And this cell - the alveolar macrophage -   has attracted increasing attention (fig.1). Macrophages are present in airways and are derived from blood monocytes and will track through the airway wall to the surface. These cells can be activated by very many different triggers and release two types of product. Inflammatory mediators, such as lipid mediators like thromboxane, leukotrienes, PAF and oxygen

P. Barnes:

radicle in the way that many other inflammatory cells can do. But they also release a whole variety of peptide mediators - cytokines - which may be of critical importance in orchestrating chronic inflammatory responses. It is likely that the pattern of cytokine release will determine the nature of the ensuing inflammatory reaction. The macrophage may therefore play a critical role in initiating the chronic inflammation of asthma, which it would do by releasing a profile of cytokines which could be specific for asthma rather than bronchiektasis or whatever other airway imflammation is possible. Of importance is that in contrast to the mast cells these cells are inhibited by steroids and are not inhibited by beta-agonists. So they have the opposite pharmacology than one would expect of a cell that should be involved in chronic inflammation.

Perhaps the most oritical cell for understanding asthma is the eosinophil because the presence of activated eosinophils in the airway is so characteristic. If we want to describe asthma today as a new disease we may call it "chronic eosinophilic bronchitis". We need to understand the role of the eosinophil in airway inflammation. Eosinophils can be activated by a number of different stimuli, but unlike the macrophages they have a rather limited repertoire of responses. They can release inflammatory mediators as other cells do, but they also release a set of basic proteins which are highly charged and these basic proteins give the eosinophil its name, because they stain with the dye eosin that makes the cell appear red. These basic proteins in some way must hold the answer to what asthma is about. These proteins are highly toxic to certain cells and were probably evolved as part of our defense against parasite invasion. They are toxic to parasites, but they also damage cells of the airway and in particular airway epipthelial cells.

In a study we looked at the effects of eosinophils on airway epithelium. Purified guinea pig eosinophils were used as they can be obtained in a high degree of purity. Under control conditions we saw a small amount of epithelial shedding with time. These eosinophils, when they are isolated, are activated to some extent. But when we specifically activate them with platelet activating factor (PAF), which is a very good activator of eosinophils, this shedding is markedly accelerated, and by 14 hours all of the epithelium has been shed completely. This may be relevant to asthma since airway epithelial shedding is characteristically found particularly in more severe asthmatics, and clumps of epithelial cells in the sputum have been used as a diagnostic feature of asthma for many years.

Eosinophils should not normally be present in the airway and the eosinophils that are present in the airway have to arise from the circulation. This occurs as a result of carefully coordinated chemical signals (fig.2). It is of critical importance to understand what these signals are, because this will tell us something fundamental about the pathophysiology of asthma. Some specific chemical signals must make the eosinophils selectively migrate into the airways and become activated. Eosinophils in order to reach the tissue have to adhere to endothelial cells in the circulation. Without this adherence they are not able to migrate into the tissue. Subsequently they move into the tissue under the influence of certain chemical signals. Then once in the tissue they can survive under the influence of other chemicals and can become activated. The signals that produce this eosinophilic information are now better understood. The most important signals are released from macrophages and T-lymphocytes. The macrophages can release PAF and interleukin-1 and tumor necrosis factor, whereas T-lymphocytes can release interleukin-3, interleukin-5 and tumor necrosis factors. All of which lead to the adherence of eosinophils, to the endothelial cells. They also migrate into the tissue under the influence of these mediators, and then become activated by mediators like PAF, interleukin-5, and granulocyte macrophage colony stimulating factor (GMCSF). These last two mediators which are cytokines probably do not activate the eosinophils directly but prime thom to respond to PAF and other stimuli. In some way they change the biochemistry of the eosinophils to allow to respond. So these signals produced by macrophages and T-lymphocytes appear to be important in leading to this eosinophilic inflammation.

So the T-lymphocyte has now taken on a rather central role in our understanding of asthma (fig. 3). We should probably regard asthma as an immunological condition, because the T-lymphocyte is a cell with a long memory, and in some way allows the persistence of the eosinophilic inflammation. We know that once asthma is set up in people, it very rarely spontaneously resolves. The T-lymphocyte is able to release these cytokines which are important to the eosinophilic inflammation, and, in order to do this, is in some way programmed. The mechanisms by which the T-lymphocyte is programmed is now being better understood and seems to involve a cell in the epithelium, the dendritic cell, which is a specialized macrophage which in some way is able to interact with antigen such as house dust mite-antigen and then migrates to regional lymph nodes. There it codes the T-lymphocytes to selectively synthesize the cytokines that are critical to eosinophilic inflammation which seem likely to be interleukin-5 and the GMCSF. So the T-lymphocyte may be extremely important.

Now we regard asthma as a complex interaction of many inflammatory cells which are all interacting with each other, and the signals they use to interact are now better understood.

The inflammatory cells produce a whole variety of mediators which then produce the pathophysiology of asthma, and it is likely that in different asthmatics at different times a different combination of mediators will be produced.

Some of these mediators are now recognized (fig. 4). They have multiple effects on the airways such as histamine, which produces bronchial constriction, mucosecretion and microvascular leakage. Histamine can be completely blocked by terfenadine, and yet terfenadine is not of value in the treatment of asthma. The obvious explanation for this is that histamine is only one of many mediators, and if you only block one, it would be unlikely that you would have a major therapeutic effect.

However some mediators may be more important than others, and there is now increasing evidence that leukotrienes particularly leukotriene D4 is of particular importance in human asthma, since potent leukotriene antagonists appear to be very beneficial in challenge studies. Another mediator which may have an important role in asthma is the platelet activating factor (PAF).

PAF closely mimics the pathophysiology of asthma and produces many of the features that we see in asthmatic inflammation, but of particular relevance is that it leads to eosinophilic inflammation in the airways when given to animals, and can also lead to increased airway hyperresponsiveness, which can be prolonged, in every species in which it has been tested including humans. So PAF can be important in the pathophysiology of airway inflammation since it may attract eosinophils, but also activates eosinophils to produce these basic proteins which can damage the airway epithelium. This effective PAF can be potentiated by cytokines like interleukin-5.

And so we should regard oytokines as playing a critical role in orchestrating asthma. The T-lymphocyte and the macrophage are the major source of cytokines, but mast cells now also been found to produce them. The important cytokines are interleukin-1, which communicates between macrophages and lymphycytes, interleukin-4, which programs T-lymphocytes to make IgE which underlies most of asthma, and these interleukins that are important to activate endothelial cells and eosinophils to lead to the characteristic inflammation.

Asthma is a complex inflammatory condition involving many inflammatory cells which in some way communicate with each other. Since this inflammation is something unique to asthma, it is not found in other airway inflammatory conditions. It now seems that cytokines play a critical role in understanding why asthmatic inflammation is initiated and more importantly, why it persists. This is clinically relevant because the effect of steroids in asthma is just to prevent the transcription of cytokines. It would seem very likely that steroids when given to patients with asthma stop the cytokines and therefore stop this chronic inflammatory process.

Discussion

Zwick: You emphasized the central role of the T-lymphocyte. Does that give us a further insight into the genetic factors of asthma?

Barnes: Progress has been made in understanding the genetic factors in asthma by identifying that atopy appears to be inherited dominantly, and the gene is localized to chromosome 11. But as 1/3 of the population are atopic, it is unlikely to give us much greater understanding about asthma.

Zach: You showed the usual concept - if I may say so - of an allergic immediate reaction, late reaction, increase of bronchial hyperresponsiveness. This seems to be all very clear for the inhalation of an allergen. How about a similar sequence after exercise, and how about viral infections?

Barnes: I did not really have time to go into this. This model of the early and the late response may be misleading in terms of understanding chronic asthma because asthmatic patients are usually not exposed to the high concentration of allergen that are given to induce these responses. I think that we should look at asthma as a subacute inflammation that follows low level chronic exposure for many years with allergen. This is quite different from giving a massive dose of allergen under laboratory conditions. It is likely that this sort of inflammation that ensues after chronic exposure is quite different from that seen in the acute late response and reactivity. Although allergen challenge has been used to investigate antiasthma drugs, I think that we now recognize that this model may not predict which drugs will be useful in controlling chronic asthma.

Exercise induced asthma can only occur with the inflammation of asthma. It is impossible to induce exercise induced asthma in normal people under any conditions. What we should be asking is why asthmatic inflammation makes people develop wheezing when they exercise. The sequence of events in exercise induced asthma is simple, and probably leads to mast cell degranulation, which may be triggered either by thermal effects or by osmotic effects and perhaps a combination of both.

Popp: Sometimes one finds eosinophils in the tissue without damage. The most important thing is the activation of the eosinophils. Is there any possibility to assess the activation of eosinophils for example in blood or in tissue?

Barnes: I think there is a possibility of studying it by

measuring eosinophil cationic proteins in the plasma. But the studies that have been done so far show a rather poor correlation with the clinical features of asthma. If you measure eosinophil products in the bronchoalveolar lavage or in sputum, that might be much more accurate.

Kummer: Just recently Dr.Venge from Upsala exited us by his data on ECP. They found an elevated level of ECP in the serum of exercise induced asthmatics, that were non-atopics. They also found increased nasal lavage ECP in pollinotics in wintertime, when there were no pollen around. There are so many people with intrinsic asthma having a lot of eosinophils and IgE, and are not allergic at all. So my question is: could you picture that atopy is just a preparedness of the organism to be sensitized which might not necessarily take place, and on the other side is there a T-cell mediated eosinophil cationic protein mechanism responsible for the great mass of our intrinsic asthmatics?

Barnes: 25 to 30 percent of the population are atopic, and they can make IgE, but only 5% have asthma. So clearly you have to have more than atopy to have asthma, but what it is to make some atopics asthmatics is not completely clear. On the other hand there are people who have asthma that are not obviously atopic. Some, such as Ben Burrows, would argue that some of those had been atopic but had lost the atopy by the time they had asthma. I think that some of them may never have had any atopy. What is striking is that the inflammation that you see in intrinsic and extrinsic asthma is very similar to eosinophil inflammation in both, and both respond to steroids very effectively. The other question relates to ECP as a diagnostic test for asthma. It may prove to be a useful measurement of asthmatic inflammation and response to therapy. If it also goes up with rhinitis, it becomes much less useful as a specific indicator for asthma, but very much research is needed.

Hargreave: I just wanted to take an issue about your comment on the relevance of the allergen inhalation model. Certainly what you say in terms that dose in laboratory being particularly high might be true when one is doing tests with, say pollen or dust mite. But in the case of certain allergens, for example pets or certain occupational allergens, all you are doing is giving a dose which in the normal situation would do exactly the same thing.

Barnes: Yes, I agree with that, of course. I mean especially with cats where people may be exposed to high doses in their home. What I wonder is whether the pharmacology of an acute induced inflammation is the same as in chronic subacute inflammation. The cellular components and the mediators may be different. We are looking for drugs that block the acute allergen induced responses, and drugs such as leukotriene antagonists may be useful. But in many chronic inflammatory responses we may need cytokine blockers.

Airway Hyperresponsiveness and Asthma

Frederick E. Hargreave

Firestone Regional Chest & Allergy Unit, St.Joseph's
Hospital - McMaster University, Hamilton, Ontario
(Canada)

Measurements of airway responsiveness have improved our understanding of asthma considerably. But of course we still have a lot to learn. In this presentation, I am going to start by saying a little about measurement before going on to talk about the interrelationships between hyperresponsiveness and the other characteristics of asthma. The objective of the presentation is to illustrate how these studies have improved our understanding of the nature of asthma and its treatment.

Measurement

Airway responsiveness can be measured by a number of agents which act through different mechanisms (1). Histamine and methacholine for example, chiefly act directly on the airways smooth muscle, while other stimuli such as exercise, hyperventilation and adenosine monophosphate, are considered to act indirectly through the release of mediators which then act on the smooth muscle. It has been suggested that these indirect stimuli may be more specific for asthma, but this has yet to be established. Up to now, airway responsiveness has been measured mainly with histamine or methacholine in a dose-response fashion.

The histamine or methacholine dose-response curve in normal subjects is shifted to the right (2,3). A larger dose of the agent is required to cause bronchoconstriction; in addition, the maximal degree of constriction possible is limited to a relatively mild degree and this produces a maximal response plateau. As airway responsiveness becomes increased, the dose-response curve moves to the left. Bronchoconstriction occurs after a low dose and the degree of constriction possible is increased so that the maximal response plateau becomes higher and, eventually, lost. The degree of airway responsiveness is usually identified by the position of the curve, and is expressed as the provocation concentration (or dose) to cause a fall in FEV1 of 20% (PC20 or PD20).

Details of the methods of measurement of histamine or methacholine airway responsiveness and their regulation have been well described (4). In the method we use, hyperresponsiveness is indicated by a PC20 of 8 mg/ml or less. The method is now published in detail (5).

Airway hyperresponsiveness is one of the characteristics of asthma, the others being airway inflammation, variable airway obstruction and symptoms. We now believe that airway hyperresponsiveness is initiated by airway inflammation.

Airway inflammation

Evidence that airway inflammation is the cause of airway hyperresponsiveness in asthma is derived from studies of direct examination of the inflammation using bronchial biopsies, bronchoalveolar lavage (BAL) and sputum, measurements of airway responsiveness in relation to laboratory exposure to allergen or chemical sensitizers to which the person is sensitized, and examination of the effects of treatment with antiinflammatory drugs on responsiveness and inflammatory indices. The direct studies using biopsies and BAL have been few in number because of their invasive nature and potential risk (6). However, they have shown evidence of infiltration with inflammatory cells, particularly eosinophils and metachromatic cells (mast cells or basophils) even in mild asthma.

In an attempt to investigate aspects of the inflammation more readily, we have examined the use of sputum, produced spontaneously and induced by inhalation of hypertonic saline, and the use of peripheral blood. The initial studies have been promising. Reproducible total cell counts and differential cell counts can be obtained in spontaneous sputum between plugs from the same specimen and different specimens on consecutive days (7). The characterics of the cell counts are different between diseases of different pathogenesis, specifically between asthma in exacerbation and smoker's unobstructive chronic bronchitis. Sputum can be induced with hypertonic saline successfully and safely (8). The induced sputum can show differences in cell counts between normal and mild asthmatics, not dissimilar to those observed in other studies using BAL. Peripheral blood counts of the progenitors for eosinophils and basophils are increased at the time of an exacerbation of asthma and fall after treatment with inhaled corticosteroid (9). These precursors come from the bone marrow possibly as a result of the release of growth and differentiation factors from the airways. These results support the further development of these methods to examine the airway inflammation noninvasively.

Evidence that airway hyperresponsiveness develops secondary to inflammation was obtained from serial measurements of responsiveness in relation to allergen and, subsequently, occupational exposure in allergic (sensitized) subjects (10). Allergen inhalation tests in the laboratory most often cause an isolated early or an early followed by late asthmatic response. The latter, but not the former, also tend to be associated with an increase in histamine or methacholine airway responsiveness when this is measured 24 hours or so after allergen inhalation. BAL performed 8 hours after exposure demonstrates an increase in eosinophils when a late response has occured but not when it has been absent. In other words, the late asthmatic response is associated with inflammation as indicated by the increase in inflammatory cells and with the development of airway hyperresponsiveness.

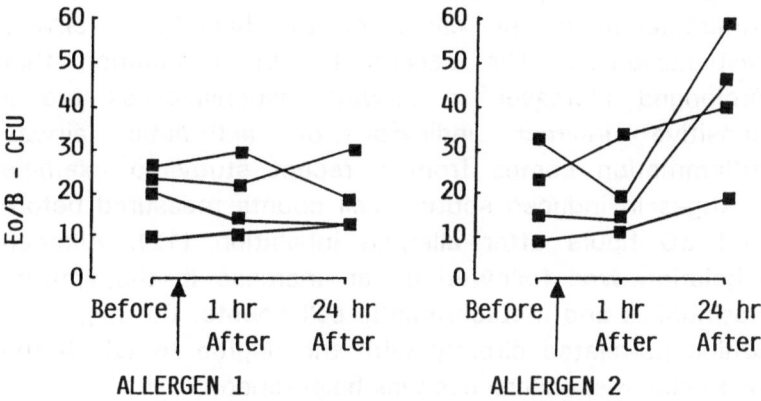

Fig. 1

Such prolonged heightening of airway responsiveness appears to be a sensitive indirect indicator of increases in airway inflammation. The evidence for this comes from two recent observations. The first is illustrated in Figure 1 which shows eosinophil and basophil progenitors in peripheral blood before, 1 hour and 24 hours after the inhalation of allergen (11). The left-hand panel shows the results in four early responses, while the right panel shows the results in the same subjects after the inhalation of a different allergen which also produced an increase in responsiveness to histamine after 24 hours. As you can see, when the allergen caused a heightening of histamine responsiveness, there was also an increase in the progenitors, but not when it did not do this. The increase in progenitors provide indirect evidence that airway inflammation is the cause of the heightened airway responsiveness. The second bit of information that prolonged increases in airway responsiveness are a sensitive indirect indicator of asthmatic airway inflammation comes from a recent study to examine changes in induced sputum cell counts measured before and 30 hours after allergen inhalation (12). Allergen inhalation was followed by an increase in the sputum eosinophils and metachromatic cell counts, the degree of which correlated directly with the degree to which the histamine responsiveness was heightened.

Variable airway obstruction

The presence of airway hyperresponsiveness is a sensitive indicator of abnormal airway function and is more sensitive than FEV1 measured at a clinic visit during the day (13).

Figure 2 illustrates this point. It shows PC20 histamine in relation to the baseline FEV1 in a group of normal subjects, and a group of past or current asthmatics. You can see that the PC20 histamine is less than 8 mg/ml indicating airway hyperresponsiveness, before the FEV1 is reduced; in fact, the FEV1 remains normal until there is a moderate to severe reduction in PC20.

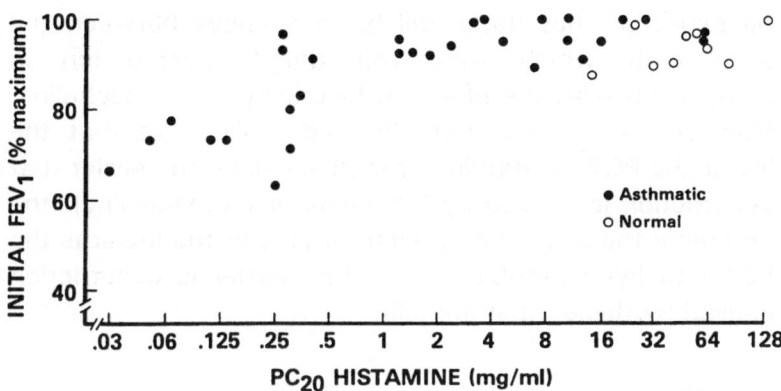

Fig. 2

Figure 3 illustrates that the degree of airway hyperresponsiveness relates closely to the degree of variable airway obstruction (13). The figure shows PC20 histamine in relation to the degree of diurnal variation of peak flow rates (PFR), obtained from daily measurements on waking and at 6:00 p.m. before and after sulbutamol over one week, expressed as a percentage of the mean of the lowest minus the highest values divided by the highest. The lower the PC20 (the greater the increase in responsiveness), the greater is the diurnal variation of flow rates (the greater is the variable airway obstruction). This relationship to variable airway obstruction is also illustrated by the ease with which bronchoconstriction can be triggered by other stimuli, such as exercise or hyperventilation of cold dry air. Each of these other stimuli act through different and specific mechanisms and it is to be expected that there will be differences between the ease with which they will trigger constriction in comparison with the effect of histamine or methacholine. However, some (but not all) studies illustrate that the lower the PC20 histamine or methacholine (the easier that constriction is caused by histamine or methacholine), the greater is the response to exercise (14) or the lower is the PD10 to hyperventilation (15) (the easier is constriction caused by these other stimuli).

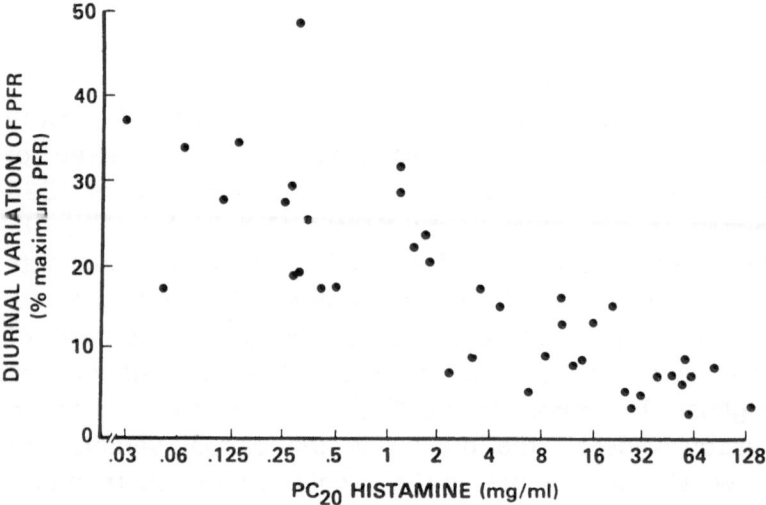

Fig. 3

I have illustrated so far that an increase in airway responsiveness and an increase in variable airway obstruction occurs in asthma. However, neither of these features are specific for asthma as illustrated by the following two points. Figure 4 shows the relationship between PC20 methacholine and the FEV1/VC ratio in a group of smokers with chronic cough and sputum consistent with chronic bronchitis (16). We did not consider that they had asthma. However, you can see that once they had developed chronic airflow limitation with an FEV1/VC ratio of less than about 70%, the PC20 was reduced into the hyperesponsive range. Many of them also had an increase in the diurnal variation of PFR (17). The results illustrate that when there is airway obstruction from another cause, that neither airway hyperresponsiveness to metacholine nor variable airway obstruction are specific for asthma; they can occur in diseases with a different pathogenesis.

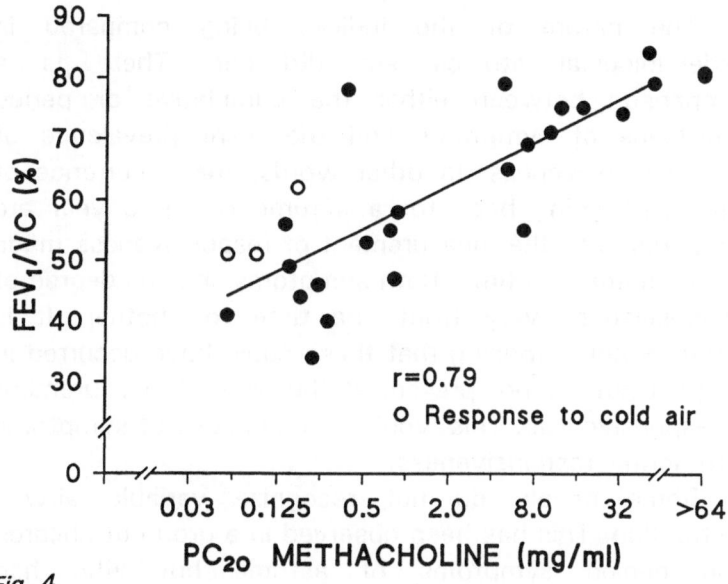

Fig. 4

Symptoms

Epidemiological studies have illustrated the discrepancies between symptoms consistent with asthma and the presence of histamine or methacholine airway hyperresponsiveness (18,19). Some subjects are hyperresponsive and have never had symptoms, while others report symptoms but have normal airway responsiveness. There are several reasons for these discrepancies which include the nature of the indices being compared, failure to recognize variable airway obstruction, the strength of the stimulus causing symptoms, and the specificity of symptoms with respect to variable airway obstruction or asthmatic airway inflammation (20).

The nature of the indices being compared in epidemiological studies are different. There is a comparison between either the cumulative or period prevalence of symptoms and the point prevalence of hyperresponsiveness. In other words, the occurrence of symptoms going back for a lifetime or for a year are compared with the measurement of responsiveness made at one moment in time. Both symptoms and the degree of responsiveness vary from one time to another. It is therefore not surprising that these could have occurred in the past but are not present at the time of measurement of responsiveness. This would be one cause of symptoms with normal responsiveness.

Some people do not recognize variable airway obstruction. This has been observed in a group of children who denied symptoms of asthma but who had methacholine hyperresponsiveness. Their diurnal variation of PFR was compared with another group of children who had symptoms and were matched for age and the degree of hyperresponsiveness, and with a third group of the same age who had no symptoms and normal responsiveness (21). The asymptomatic hyperresponsive children had the same degree of variable airway obstruction as the symptomatic hyperresponsive group, and were different to the children with normal responsiveness. The hyperresponsiveness in the asymptomatic children therefore suggests an abnormality of airway function, the cause of which has yet to be determined.

Symptoms and variable airway obstruction can be produced in subjects with normal histamine or methacholine responsiveness when the stimulus for bronchoconstriction is strong. This can occur when the stimulus is exposure to an allergen or chemical sensitizer which, presumably, causes enough mediator release to cause airway constriction, just as histamine or methacholine will do in many normal subjects if the dose is high enough.

Finally, symptoms are not necessarily specific for variable airway obstruction (and can presumably be due to a variety of other causes), and they can be produced by asthmatic airway inflammation when airway hyperresponsiveness and variable airway obstruction are absent. The former was illustrated in a clinical study of 51 consecutive patients referred to our chest clinic (23). All of them had current symptoms of asthma and normal spirometry. Methacholine hyperresponsiveness was only

present in a proportion of them; in other words, there was
a discrepancy in many between the presence of current
symptoms consistent with asthma and an absence of
hyperresponsiveness (and, from what I have already said,
presumably an absence of variable airway obstruction).
Furthermore, the physician was not able to predict
whether hyperresponsiveness was likely to be present, he
or she being wrong 40% of the time. Presumably, such
symptoms in people with normal responsiveness could
have a number of causes. One of them, recently
recognized, is airway inflammation with asthmatic
characteristics which has not also produced
hyperresponsiveness or variable airway obstruction (24).
This is illustrated in Fig. 5, which shows sputum cell
counts of eosinophils and metachromatic cells in three
groups of subjects, one of asthma in exacerbation (●), a
second of smokers with non-obstructive bronchitis (.) and
a third of patients presenting with a chronic productive
cough (■) which, like the asthmatics in exacerbation, was
reversed by treatment with inhaled corticosteroid. This
third group had normal spirometry and normal airway
responsiveness but they had an increase in eosinophils
and metachromatic cells in the sputum, similar to that
present in the asthmatic group.

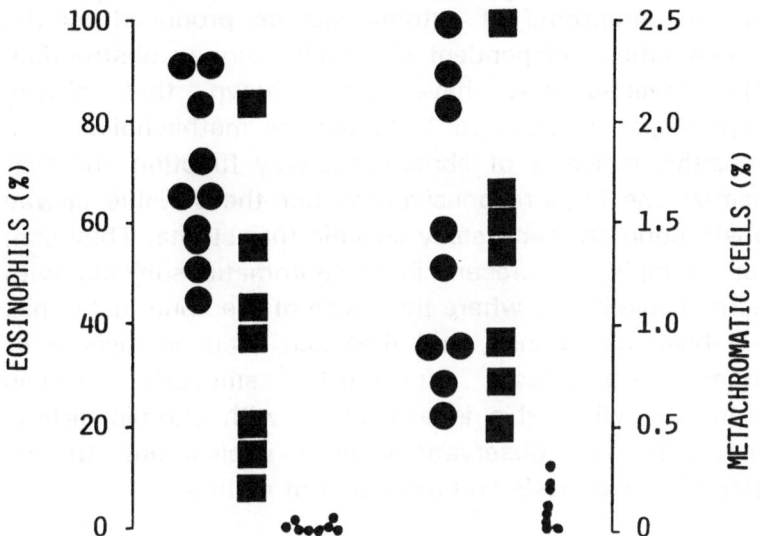

Fig. 5

Conclusion

In summary, studies of the interrelationships between airway hyperresponsiveness and other characteristics of asthma have advanced our understanding of asthma. They have provided evidence that airway hyperresponsiveness, variable airway obstruction and symptoms are secondary to airway inflammation (and/or its effects), changes in airway hyperresponsiveness are a sensitive indirect indicator of changes in airway inflammation, the degree of airway hyperresponsiveness is probably a determinant of the degree of variable airway obstruction, and that some of the symptoms of asthma can be produced by the inflammation independent of variable airway obstruction. The investigations have also shown that airway hyperresponsiveness to histamine or methacholine is a sensitive indicator of abnormal airway function and that neither the hyperresponsiveness nor the variable airway obstruction are necessarily specific for asthma. They can, for example, be present in asymptomatic subjects with normal spirometry where the cause of the abnormality has not been determined, and they can occur in disease of other pathogenesis, specifically smoker's chronic bronchitis when this is associated with chronic airflow limitation. The observations have implications to the definition, diagnosis and treatment of asthma.

Discussion:

Zach: I would like to ask another type of question relating to the correlation of symptoms on one side and measured variables on the other. I think one fact that you did not mention is the perception of symptoms by the patient which varies from patient to patient enormously. Because usually what we call symptoms are patient reported symptoms. I think we should have a lot of variation when patients report the same type of symptoms or the same type of airway obstruction in a very different way.

Hargreave: Yes, you are correct. This is a cause of discrepancy between symptoms and airway hyperresponsiveness. The patient may not recognize airway constriction (and so have no symptoms) or may not recognize symptoms which are present. The latter can be recognized by asking if symptoms occur at the time of the histamine or methacholine test and whether these have ever occurred before. Some who have denied symptoms before will then have symptoms during the test which they then realize to have actually occurred in the past.

Popp: May I ask you for which purpose you propose self assessment by peak flow recording. Only for assessing hyperresponsiveness for diagnostic approach or for therapeutic control also?

Hargreave: I do not use daily measurements of peak flow rate (PFR) at home to confirm the diagnosis of asthma (of variable airway obstruction). This is because there is a dose relationship between the degree of daily variability of PFR and the degree of histamine or methacholine airway responsiveness and the latter measurement can be made very quickly and reliably. PFR measurements at home are, however, useful to investigate the diagnosis of occupational asthma (although serial measurements of airway responsiveness are also useful for this purpose), and to help with sorting out the treatment of more severe asthma.

Kummer: If the patient is very apprehensive and very concerned about the symptoms and experiences his dyspnea in a very dramatic way, he would take a lot of medication. But if it is a more slobby or indolent patient, then he would not take medication carefully, he would rather run into trouble and come to the hospital in emergency. Dr. Holgate's group was not able to correlate the severity of bronchial hyperresponsiveness with the medication intake. The more anxious patients are better off actually than the indolent patients. Would you feel the missing link could be the compliance of the patients taking the medication?

Hargreave: I agree that one reason for the discrepancies he described between changes in symptoms, variability of flow rates and PC20 may be differences between patients in their ability to recognize airway obstruction. However, I believe that there are other reasons which relate to study design, accuracy of performance of measurements, timing of measurements and the condition causing the symptoms. For example, the measurement of PC20 was only taken to the cut point between normal and abnormal and several subjects had a normal value at times. They could have an exacerbation of symptoms and an increase of responsiveness within the normal range without this being detected. The accuracy of PFR is influenced by the depth of inspiration before the measurement; a cough preventing maximal inspiration might lower the PFR in the absence of airway obstruction.

Kerrebijn: Would both speakers agree that peak flow variability or PC20 or whatever is not a quantitative but only a qualitative indicator of disease activity? I think that is the message which Holgate gave, and I think that is true.

Hargreave: I think within subjects it is a quantitative indicator, between subjects it may not be.

Kerrebijn: I am not sure it is a quantitative indicator within subjects. Do you consider your normal subjects with an increased responsiveness being just normal or do you think that there is something wrong in their airways?

Hargreave: I believe the majority of them has got something wrong in their airways. In some of them it may be inflammation indicated by an infiltration of cells. In others it might be structural changes.

Kerrebijn: So it indicates that there is something wrong, but PC20 does not quantify the amount of "wrongness" so to say. They have no symptoms and probably their peak flow variability is not highly increased.

Hargreave: Yes, I agree.

Acknowledgements

The work reported in this presentation has been particularly supported by grants from the Medical Research Council of Canada, Astra Pharma Inc. and Boehringer Ingelheim (Canada) Ltd.

References

1. Pauwels R, Joos G, Vander Straeten M: Bronchial hyperresponsiveness is not bronchial asthma. Clin Aller (1988) 18:317-321
2. Woolcock AJ, Salome CM, Yan K: The shape of dose-response curve to histamine in asthmatic and normal subjects. Am Rev Respir Dis (1984) 130:71-75
3. Sterk PJ, Daniel EE, Zamel N, Hargreave FE: Limited bronchoconstriction to methacholine using partial flow-volume curves in nonasthmatic subjects. Am Rev Respir Dis (1985) 132:272-277
4. Hargreave FE, Woolcock AJ, eds:. Airway responsiveness: measurement and interpretation. Mississauga: Astra Pharmaceuticals Canada Ltd, 1985
5. Juniper EF, Cockroft DW, Hargreave FE: Histamine and methacholine inhalation tests: Tidal breathing method laboratory procedure and standardisation. Canadian Thoracic Society AB Draco, Lund Sweden, 1991
6. Djukanovic R, Roche WR, Wilson JW, Beasley CRW, Twentyman OP, Howarth PH, Holgate ST: Mucosal inflammation in asthma. Am Rev Respir Dis (1990) 142:134-157
7. Gibson PG, Girgis-Gabardo A, Morris MM, Mattoli S, Kay JM, Dolovich J, Denburg J, Hargreave FE: Cellular characteristics of sputum from patients with asthma and chronic bronchitis. Thorax (1989) 44:689-692

8.   Pin I, Gibson PG, Kolendowicz R, Dolovich J, Denburg J, Hargreave FE: Induced sputum cell characteristics in normal (N) and asthmatic (A) subjects. Eur Respir J (1990) 3(10):255S

9. Gibson PG, Dolovich J Girgis-Gabardo A, Moris MM, Anderson M, Hargreave FE, Denburg JA: The inflammatory response in asthma exacerbation: changes in circulating eosinophils, basophils and their progenitors. Clin Exp Allergy (1990) 20:661-668

10. O'Byrne PM, Dolovich J, Hargreave FE: Late asthmatic responses. Am Rev Respir Dis (1987) 136:740-51

11. Gibson PG, Manning PJ, O'Byrne PM, Girgis-Gabardo A, Dolovich J, Denburg JA, Hargreave FE: Allergen-induced asthmatic responses: relationship between increases in airway responsiveness and increases in circulating eosinophils, basophils and their progenitors. Am Rev Respir Dis (1991) 143:331-335

12. Pin I, Freitag AP, O'Byrne P, Girgis-Gabardo A, Denburg JA, Dolovich J, Hargreave FE: Changes in the cellular profile of induced-sputum after allergen-induced asthmatic responses. J Allergy Clin Immunol (1991) 87 (1 Pt.2): 249

13. Ryan G, Latimer KM, Dolovich J, Hargreave FE: Bronchial responsiveness to histamine: relationship to diurnal variation of peak flow rate, improvement after bronchodilator and airway caliber. Thorax (1982) 37:423-429

14. Anderton RC, Cuff MT, Frith PA, et al: Bronchial responsiveness to inhaled histamine and exercise. J Allergy Clin Immunol (1979) 63:315-320

15. O'Byrne PM, Ryan G, Morris M, McCormack D, Jones NL, Morse JLC, Hargreave FE: Asthma induced by cold air and its relation to nonspecific bronchial responsiveness to methacholine. Am Rev Respir Dis (1982) 125:281-285

16. Ramsdale EH, Morris MM, Roberts RS, Hargreave FE: Bronchial responsiveness to methacholine in chronic bronchitis: relationship to airflow obstruction and cold air responsiveness. Thorax (1984) 39:912-918

17. Ramsdale EH, Morris MM, Hargreave FE: Interpretation of the variability of peak flow rates in chronic bronchitis. Thorax (1986) 41:771-776

18. Sears MR, Jones DT, Holdaway MD, Hewitt CJ, Flannery EM, Herbison GP, Silva PA: Prevalence of bronchial reactivity to inhaled methacholine in New Zealand children. Thorax (1986) 41:283-289

19. Woolcock AJ, Peat JK, Salome CM, Yan K, Anderson CD, Schoeffel RE, McCowage G, Killalea T: Prevalence of bronchial hyperresponsiveness and asthma in a rural adult population. Thorax (1987) 42:361-368

20. Cockcroft DW, Hargreave FE: Airway hyperresponsiveness. Relevance of random population data to clinical usefulness. Am Rev Respir Dis (1990) 142:497-500

21. Gibson PG, Mattoli S, Sears MR, Dolovich J, Hargreave FE: Variable airflow obstruction in asymptomatic children with methacholine airway hyperresponsiveness. Clin Invest Med (1988) 11(4): C105

22. Hargreave FE, Ramsdale EH, Pugsley SO: Occupational asthma without bronchial hyperresponsiveness. Am Rev Respir Dis (1984) 130:513-515

23. Adelroth E, Hargreave FE, Ramsdale EH: Do physicians need objective measurements to diagnose asthma? Am Rev Respir Dis (1986) 134:704-707

24. Gibson PG, Dolovich J, Denburg J, Ramsdale EH, Hargreave FE: Chronic cough: eosinophilic bronchitis without asthma. Lancet (1989) i:1346-1348

Physiologic Correlates of Increased Airway Responsiveness

Roland H. Ingram

Jr, Hennepin County Medical Center and University of Minnesota (USA)

The airway caliber change that occurs immediately following a deep inhalation (DI) in asthmatic subjects varies. There may be a decrease in airway caliber - i.e. bronchial constriction - or there may be no change, or dilation of the airways may occur. I would like briefly to review data which support the idea that when the clinical state and the stimulus used to produce bronchial obstruction are considered, there is a pattern that might be of value clinically and epidemiologically.

The easiest way to assess the effect of a DI is by using the standard flow-volume curve. A forced expiratory maneuver from functional residual capacity to residual volume, followed by a rapid inhalation to total lung capacity before a maximal expiratory maneuver, i.e. a standard forced vital capacity maneuver. The effect of a DI on airway caliber can be assessed by comparing the maximum-to-partial flow ratio at a given volume. Maximal-partial (M/P) ratios greater than one indicate a dilating effect of a deep breath. M/P ratios of less than one indicate a constrictor effect of a deep breath.

The M/P ratios correlate reasonably well along the line of identity with specific conductance ratios before and after a deep breath if a wide range of values is used for comparison (SGAW ratio = 0.94 M/P ratio + 0.14; R = .81) (1). Therefore the M/P ratio, which is much more easily performed and measured,   can be used interchangeably with SGAW ratios. Hereafter, either will be referred to as volume history ratio (VHR). A value greater than 1.0 indicates dilation after a DI; a value less than 1.0 indicates constriction.

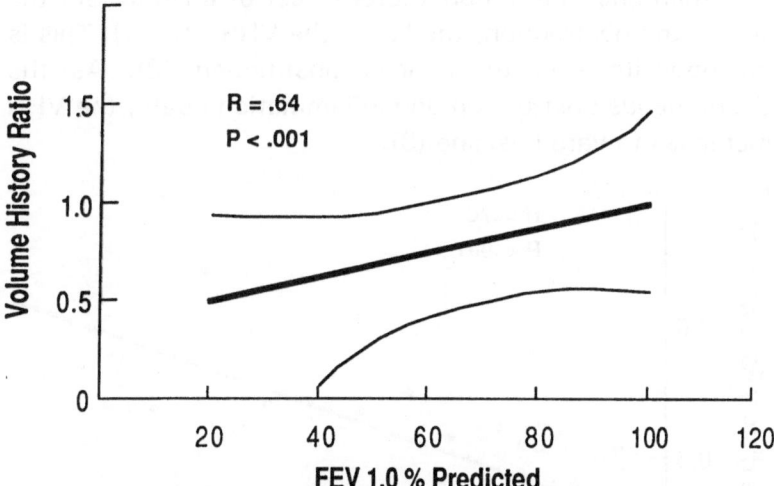

Fig. 1: Relationship between severity of obstruction in spontaneous (i.e.not induced) asthma and the VHR. The worse the spontaneous obstruction, the lower the VHR. In contrast, induced obstruction is associated with a higher (>1.0) VHR. Redrawn from reference 2.

The sign and magnitude of airway caliber change immediately following a DI in asthmatics correlates with both the site and mechanism of airway obstruction. Constriction of conducting airways, such as produced by histamine or methacholine is lessened by a DI, does not change baseline responsiveness and may be used to assess the degree of responsiveness. In fact, the greater the degree of induced obstruction, the higher the VHR.

Stimuli that are thought to cause an increase in airway responsiveness through producing inflammation of the airways (such in spontaneous asthma in association with a viral respiratory illness) are associated with VHR less than one - i.e. a constrictor effect of a DI. In fact the worse the obstruction, the lower the VHR. (fig. 1). This is an opposite sign to induced obstruction (2). As the spontaneous obstruction and inflammation abate, the VHR increases toward baseline (3).

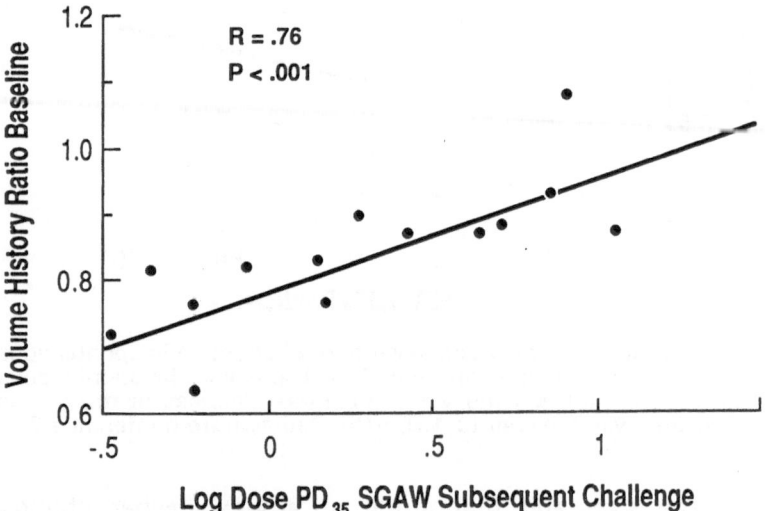

Fig. 2: VHR at baseline in mild asthmatics correlates with responsiveness (expressed as provocative dose for a 35% decrease in specific conductance) to a subsequent methacholine challenge. Redrawn from reference 4.

A constrictor or even a diminished dilator effect of a DI in asthmatics, both at baseline with normal lung function and during induced obstruction with methacholine, correlates with degree of hyperresponsiveness.

Figure 2 presents the baseline data which were taken from a group of volunteers who had totally normal lung function and were considered to be relatively mild asthmatics. None was on continuous treatment. In these mild asthmatics, only one dilated. There was a reasonable correlation, i.e. the more they constricted, following a DI at baseline, the more responsive they were to a subsequent challenge with methacholine (4).

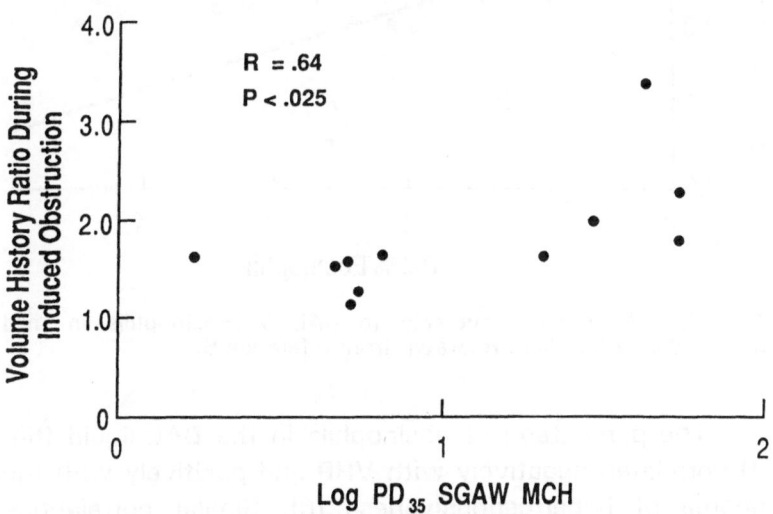

Fig. 3: VHR during obstruction in mild asthmatics correlates with responsiveness - i.e. the more responsive individuals reverse their obstruction less with a DI than ones less responsive. Redrawn from reference 5.

In another group of mild asthmatics, all increased their VHR with methacholine induced obstruction (fig. 3). However, the VHR during induced obstruction was less in the more responsive subjects. Those who were less responsive and who were given the same degree of airway obstruction had a much greater VHR (5).

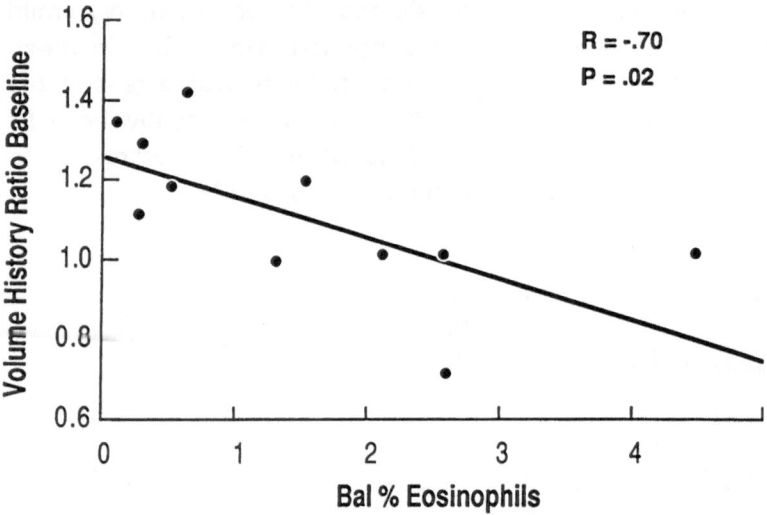

Fig. 4: VHR relates inversely to BAL % eosinophils in mild asthmatics at baseline. Redrawn from reference 6.

The percentage of eosinophils in the BAL liquid (fig. 4) correlated negatively with VHR and positively with the degree of hyperresponsiveness (6). Similar correlations were found with the BAL   total protein (indicating microvascular leak), histamine levels, and sulfidopeptide leukotriene levels.

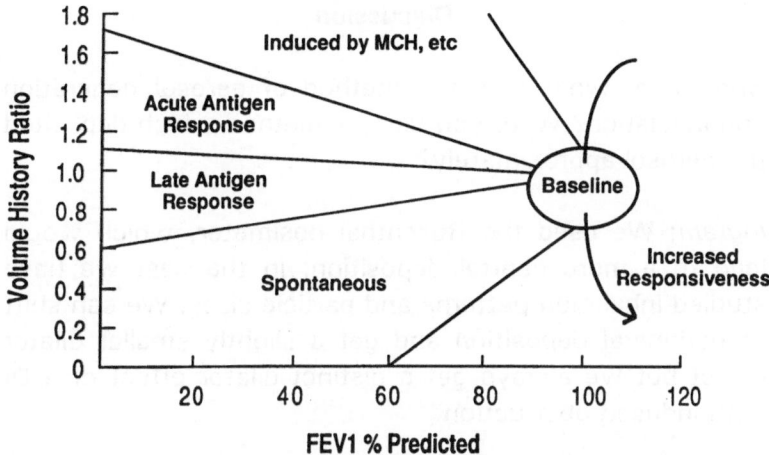

Fig. 5: VHR in relation to the kind of obstruction and its severity in asthmatics. The arrow piercing the baseline values indicates what is shown in figure 2 - i.e. greater responsiveness is found in those with a lower VHR at baseline.

Figure 5 gives a summary of the expected VHR varying from less than one with spontaneous attacks to values much greater than one with the usual stimuli used to challenge the airways.

We have inferential but compelling evidence that an imbalance between airway and parenchymal hysteresis explains this apparently disparate behavior amongst asthmatics. Time does not permit a review of that evidence but most of it is published (7, 8).

Discussion

Hargreave: What was the method of aerosol generation and inhalation? Were you using a method which deposited the aerosol approximately?

Ingram: We used the Rosenthal dosimeter, which would lead to a more central deposition. In the past we have studied inhalation patterns and particle sizes. We can shift to peripheral deposition and get a slightly smaller dilator effect but we always get a distinct dilator effect of a DI with induced obstruction.

Schinko: Is there any difference between eosinophilic and neutrophilic leukocytes in peripheral airway inflammation? You skipped the neutrophilic leukocytes.

Ingram: I don't know. We found eosinophils, and we didn't find neutrophil increases either in guinea pigs or in humans with the stimuli we used. So I would go along with Dr. Barnes, eosinophils seem to be important cells in the asthmatic. Perhaps neutrophils are, too, in other settings and in response to other stimuli.

Kummer: These forced flow-volume curves that you showed as a response to forced inspiration and then expiration were resembling the one that we usually see in small airways disease when this curve is sort of hanging through. Could you imagine that in procedures that cause reactions in the periphery like inflammation in spontaneous asthma there might be more air trapping?

Thus a mechanical component would not cause constriction but compression of the peripheral airways.

Ingram: Your point is that you had expected a difference in residual volume. Yes, you do find one and it is small. As a matter of fact, if you look at several hundreds of these, you find that the residual volume is higher on the partial curve when the VHR is greater than one, and the residual volume is lower on the partial when the VHR is less than one. The range at the maximum is about 200 ccm. So your prediction is quite correct.

Zach: Maybe I ask you a very general question for general benefit. Are you introducing something more or less like a new lung function test?

Ingram: I think it is more of a physiological analysis rather than a new test. It does not require anything new in the way of equipment. The standard spirometer will do it. And one must pay careful attention to make sure that the effects of gas compression are the same on the two curves, which you can do by assessing curve shape. There must be a brisk rise to a peak and the shape of the curve falling thereafter to a clear flow limitation configuration on both the partial and maximal flow volume curve. You can perform the test with what you have in your laboratory right now.

Zach: My question went in a slightly different direction. Would you mind speculating on to which clinical questions one could really apply this effect of a deep inhalation?

Ingram: I am not really proposing any immediate clinical utility. However, a relationship between VHR, inflammation and hyperresponsiveness would allow you to use VHR to assess changes in the degree of inflammation and hyperresponsiveness. For example, in studies of therapeutic or environmental events and their relationship to these two variables, VHR would be easy to measure. Perhaps serial studies of VHR on patients might influence your decision to start steroids or adjust their doses. It is conceivable that you could use VHR to make an early decision to admit a severe asthmatic. All of this would have to be studied, of course, to assess any real usefulness.

References

1. Ingram RH Jr. Physiological assessment of inflammation in the peripheral lung of asthmatic patients. Lung (1990) 168:237
2. Lim TK, Pride NB, Ingram RH Jr. Effects of volume history during spontaneous and acutely induced obstruction in asthma. Am Rev Resp Dis (1987) 135:591
3. Lim TK, Ang SM, Rossing TH, Ingenito EP, Ingram RHJr: The effects of deep inhalations on maximal expiratory flows during intensive treatment of spontaneous asthmatic episodes. Am Rev Resp Dis (1989) 140:340
4. Ingram RH Jr. Site and mechanism of obstruction and

hyperresponsiveness in asthma. Am Rev Respir Dis (1987) 136:(4)2:S62

5. Kariya ST, Thompson LM, Ingenito EP, Ingram RH Jr. Effect of lung volume, volume history and methacholine on tissue viscance in man. J Appl Physiol (1989) 66:977

6. Pliss LB, Ingenito EP, Ingram RH Jr. Responsiveness, inflammation and effects of deep breaths on obstruction in mild asthma. J Appl Physiol (1989) 66:2298

7. Burns CB, Taylor WR, Ingram RH Jr. Effects of deep inhalation in asthma: relative airway and paren-chymal hysteresis. J Appl Physiol (1985) 59: 1590

8. Wang YT, Thompson LM, Ingenito EP, Ingram RH Jr. Effects of increasing doses of beta agonists on airway and parenchymal hysteresis. J Appl Physiol (1990) 68:363

Bronchial Provocation Testing in Children - Methods and Clinical Relevance

Karl F. Kerrebijn

*Dean, Faculteit der Geneeskunde en
Gezondheidswetenschappen, Erasmus Universiteit,
Rotterdam (The Netherlands)*

Bronchial responsiveness is measured by means of dose-response-curves to bronchoconstricting agents. Bronchial sensitivity is determined by the position of the dose-response-curve. In asthmatics it is shifted to the left. So sensitivity is measured by the position of the curve, a leftward shift is called a hypersensitivity. But that is not the only phenomenon, there is more. Normals and mild asthmatics have a plateau, they have a maximal level of bronchoconstriction. That is very important because a maximal level of bronchoconstriction indicates that airways can only narrow to a certain level. Normals show a plateau at low level, mild asthmatics and COPD-patients show intermediate plateau values, and in severe asthmatics a plateau cannot be measured unless we bring the patient into a danger.

The maximal constriction plateau may be even more important than the position of the dose-response-curve.

Prejunctional mechanisms relate to epithelial damage or malfunction, neural control, inflammatory cell number and activity, interaction between inflammation and other structures on the airway wall and elimination of mediators and inflammatory cells. Postjunctional mechanisms relate to smooth muscle contractility and viscous and elastic loads. Important and determining maximal airway narrowing is the swelling of the airway wall, intraluminal exsudate and secretions. The prejunctional mechanisms are mainly responsible for the shift of the dose-response-curve to the left, whereas the postjunctional mechanisms are mainly responsible for the level of the maximal airway constriction, the level of the plateau.

Bronchoconstriction can be provoked by direct stimuli (methacholine and histamine etc) and indirect stimuli like exercise, cold air, osmotic challenge, SO_2 or metabisulfite, allergens etc. It may well be that the indirect stimuli which work at the prejunctional level are more important for bronchial hyperresponsiveness and asthma than the direct stimuli. So far we had most attention to the direct stimuli, and I think we should rather focus on indirect stimuli in the future.

As for methods of measuring bronchial responsiveness, I will not go into details, but I will focus on two aspects which are important for paediatrics. The first one is the administration of the aerosol. There are two methods possible. The method of tidal breathing for two minutes and the dosimeter method. In adults both gave comparable results, the relationship between the PD20 or PC20 obtained by tidal breathing and via the dosimeter is linear. They do not have exactly the same outcome, but they are very well comparable. This is not necessarily so in children with growing airways. Therefore we did a study in children aged 6 to 18 years in order to see whether both methods are comparable in growing individuals and therefore can be used for longitudinal studies.

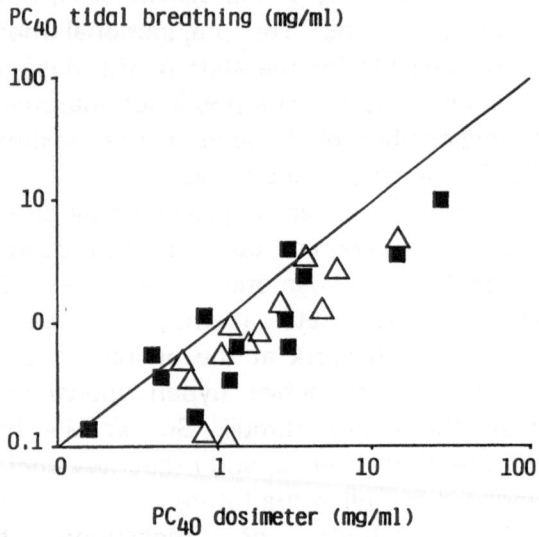

Fig. 1: Relation between values for bronchial responsiveness (PC40) to histamine obtained with the tidal breathing technique and the dosimeter technique.
o = First measurement by dosimeter; * = first measurement with tidal breathing.

On figure 1 it is shown that the ratio of PC40 obtained by tidal breathing and the ratio of PC40 obtained by the dosimeter do not differ a lot, and that the ratio of PC40 tidal breathing and PC40 dosimeter is not dependent of age (Fig. 1a).

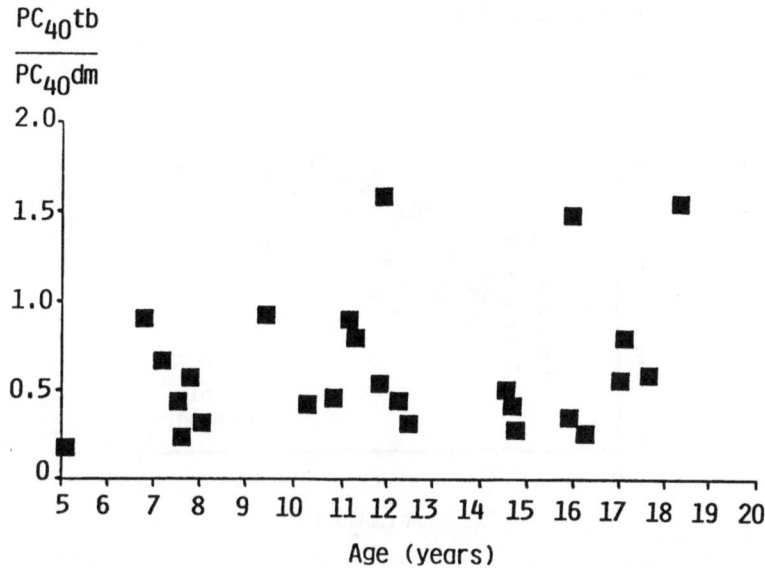

Fig. 1a: Ratio of values for bronchial responsiveness to histamine obtained with the tidal breathing technique (PC40tb) to those obtained with the dosimeter technique (PC40dm) according to age. $r = 0.27$ ($p > 0.10$).

This reassures us about the validity of both methods in growing children.

In children from the age of 6 years on the usual techniques can be applied, such as spirometry or plethysmography etc. But under the age of 6 this is not necessarily so. In our laboratory we applied two methods in younger children. The first is the forced oszillometry to measure a pulmonary resistance, the second is the measurement of transcutaneous PO_2.

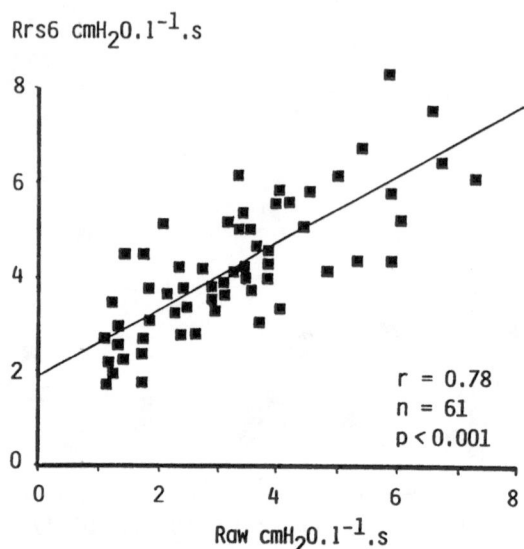

Fig. 2: Correlation of Rrs values measured by forced oscillation technique (FOT) and airway resistance (Raw) measured by whole body plethysmography.
Rrso: pulmonary resistance at 6 Hz.

Figure 2 shows that airway resistance measured by the use of a bodyplethysmograph and pulmonary resistance - here at a frequency of 6 Hz - are very well comparable with a correlation coefficient of 0,78 in 61 children. This validates the forced oscillation method as a technique for measuring airway narrowing. Airway responsiveness measured by the forced oscillation technique and by FEV1 are also very well comparable as shown in figure 3.

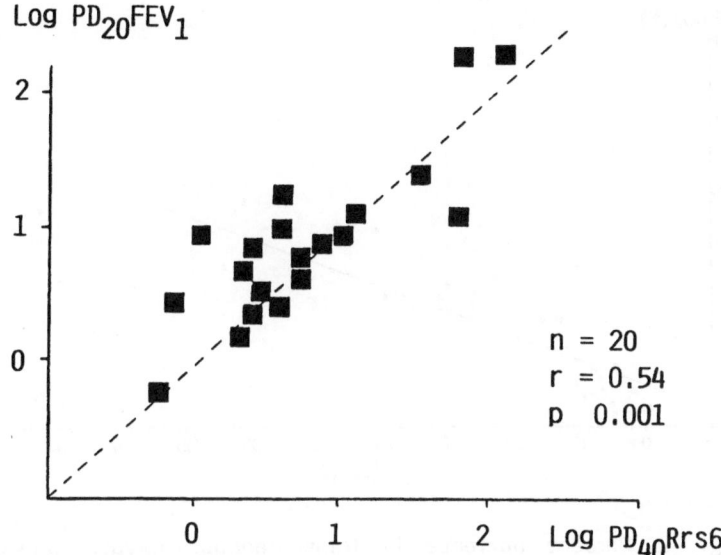

Fig. 3: Bronchial responsiveness to methacholine by PD20FEV1 and PD40Rrso.
PD20FEV1: provocative dose of agonist producing a 20% fall in forced expiratory volume in one second from baseline; PD40Rrso: provocative dose of agonist producing a 40% increase in pulmonary resistance from baseline.

This again validates the forced oscillation method to be used for measuring airway responsiveness, in this case to methacholine.

Forced oscillometry is not so easy to apply in younger children. So we compared the change in transcutaneous oxygen tension with the change in FEV1, and the results are shown in figure 4.

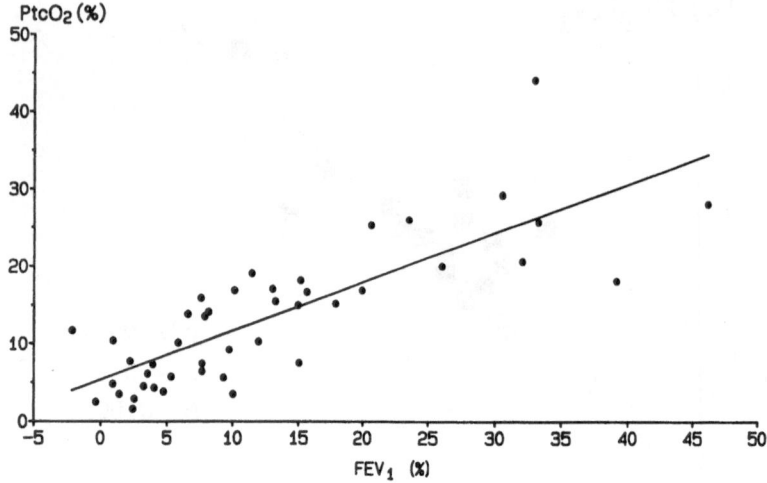

Fig. 4: Percent decrease in transcutaneous oxygen tension (PTCO$_2$) and forced expiratory volume in 1 second (FEV1) at dose step 4 (1 g. l^{-1}). The curve depicts the least squares fitted regression line (r = 0.81; p<0.001).

We repeated both measurements on two separate days and it is obvious that the method is quite reproducable (Fig. 4a). So, we use the transcutaneous PO$_2$-measurement now in children from the age of 2 1/2 to 3 years onwards.

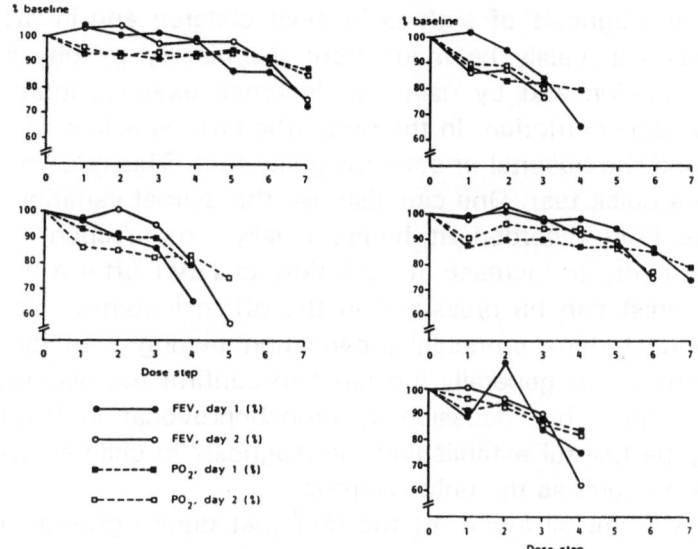

Fig. 4a: Within - subject reproducibility of PTCO$_2$ and FEV1 dose-response curves. For abbreviations see figure 4.

Do we need bronchial provocation testing in children, what are the indications?

Possible Indications:

* diagnosis of asthma
* confirmation of the clinical relevance of atopy
* marker of disease activity
* guideline for treatment confirmation of "cure" in an asymptomatic patient
* research

A diagnosis of asthma in most children and in most adults can easily be made from the history by physical examination and by using for instance exercise induced bronchoconstriction. In the office the child is asked to run around the hospital or something like that. This is a cheap and a quick test. One can also use the diurnal variation of peak flow recorded at home. Finally the response to treatment, an increase in peak flow or FEV1 after a beta-2-agonist can be measured in the office instantly. So in general I think bronchial provocation testing with direct stimuli is not generally indicated to confirm the diagnosis of asthma, but occasionally bronchoprovocation testing may be helpful establishing the diagnosis in children who report cough as the only symptom.

We are all aware of the fact that during growth the number of children who become atopic is increasing. Now the question arises, what the clinical significance of positive skin tests and of the presence of antigen-specific IgE in serum is for patients who have rhinitis or symptoms of asthma. Do we need bronchoprovocation testing with allergens to confirm the clinical relevance of atopy?

We think that clinically relevant atopy is likely if antigen specific IgE occur - measured my means of skin testing or found in the blood - plus the presence of non-specific bronchial hyperresponsiveness. From earlier studies we know that atopy itself does not cause symptoms of asthma. Symptoms occur only when the patient has also hyperresponsive airways. Most asthmatic children have hyperresponsive airways, and we do not consider bronchial provocation testing with allergens indicated in patients, but only for research purposes. Nonspecific bronchial provocation testing, but not provocation with allergens, may therefore be indicated in patients in whom the clinical relevance of atopy is to be confirmed.

The next point and the most important one: can we use bronchial provocation testing as a guideline for treatment and for confirming a remission in an asymptomatic patient?

Patients and most doctors find symptoms the best guideline for treatment. Peak flow variability both within and between days can be helpful in patients with severe asthma and in those suspicious for over- and underestimation. Secondly the position and maximal response plateau of the dose-response-curve can be used for instance in patients suspicious of over- and underdiagnosis, more importantly in patients who consider themselves cured. That is often the case in adolescence. Remission ("cure") is defined as a lack of symptoms, normal airway caliber, litte increase in FEV1 or peak expiratory flow (PEF) after bronchodilator and bronchial responsiveness within the normal range.

But first we have to answer the question: is bronchial hyperresponsiveness a sensitive marker of pathology, of disease activity?

Bronchial hyperresponsiveness is closely associated with airway inflammation. Bronchial hyperresponsiveness and the number of mastcells, histamine levels and levels of other mediators in BAL-fluid correlate positively. The same is true for biopsy studies, influx and activation of granulocytes (mostly eosinophils) and epithelial damage or the number of epithelial cells.

There is an association between bronchial hyperresponsiveness and symptoms in groups of patients, but this is not necessarily the case in individual patients. Some report a lot of symptoms but that does often not correlate with changes in daily peak-flow, and it does not correlate well with bronchial hyperresponsiveness either. I think, most of us have similar experiences. So we can conclude that in general bronchial hyperresponsiveness reflects the presence of airway pathology, and airway pathology can be either active inflammation or residual pathology like smooth muscle hypertrophy, changes in elastic load etc. In individual patients bronchial hyperresponsiveness does not necessarily reflect the severity of airway pathology and is therefore not necessarily correlated well with the presence of symptoms.

We can use bronchial hyperresponsiveness as a marker of pathology, but as a qualitative and not as a quantitative one. The role of indirect stimuli for the measurement of bronchial responsiveness has not been established. I think that for reasons I have mentioned before responsiveness to indirect stimuli may better reflect the state of pathology. We need further studies on this aspect.

The next question arises. How long should we continue treatment?

We know that we can diminish bronchial responsiveness for instance with inhaled corticosteroids. Should we continue to do that until airway caliber and bronchial responsiveness are within the normal range or are we satisfied if the patient reports that he is free of symptoms?

That is important because a recent study published in American Review of Respiratory Disease has shown that airway caliber (FEV1 % predicted) as well as the sensitivity to histamine in childhocd are predictors of asthma at the age between 25 and 30 years. If we suppress bronchial hyperresponsiveness, will that change the prevalence of adulthood asthma?

Currently we are doing a longitudinal study comparing the effects of an inhaled corticosteroid together with a beta-2-agonist to beta-2-agonist alone. Up to now 24 months of follow-up have been completed, and figure 5 shows that in the corticosteroid group FEV1 increased from about 77% predicted baseline after about 8 months to 85%. So it did not normalize, but after about 8 months a plateau was reached and nothing changed later on. In the group with beta-agonist solely nothing happened. FEV1 even dropped somewhat. That is in accordance with previous observations.

So obviously longtime treatment with antiinflammatory medication will improve airway caliber. It will also improve bronchial hyperresponsiveness. But in contrast to airway caliber there is not a plateau. In the majority of patients a plateau was not reached after nearly two years of treatment. So this indicates that airway caliber and bronchial hyperresponsiveness are rather independent factors. Even after longterm treatment there is an increase, the level reached is about three dose steps higher than baseline. But it has only reached normality in about one fourth of the patients after a mean treatment period of 22 months. After 12 months of treatmeant about 60% of patients were free of symptoms.

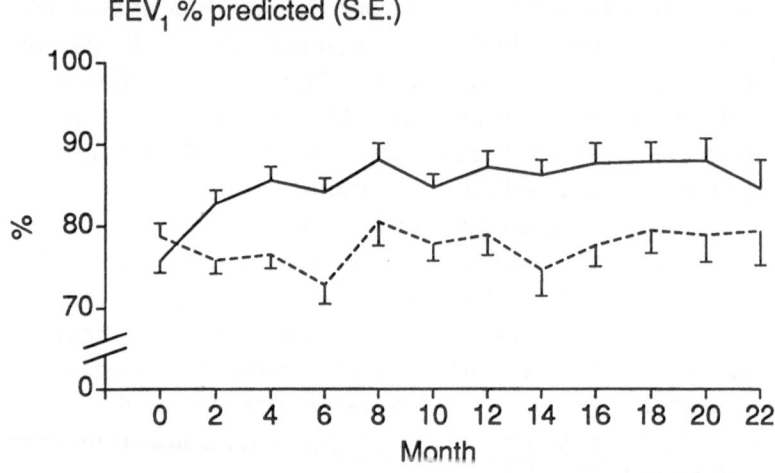

Fig. 5: see text for details

I think that we can conclude from what we have seen so far that if a patient is free of symptoms, and if airway caliber and airway responsiveness are within the normal range, there is a remission. I do not think that there will be much controversy about that. Whether a remission means a cure, I do not know; nobody has ever shown that, and that is an interesting question. The second point is that airway hyperresponsiveness indicates pathology, active or residual, we cannot separate these two for the time being. The degree of airway hyperresponsiveness and the degree and kind of pathology are not closely related and in individual patients pathology is not necessarily related to symptoms.

Two interesting points come up while we speculate upon the value of suppression of airway hyperresponsiveness: longterm outcome is unknown, and the value of indirect stimuli as an indicator of pathology and predictor of outcome is also unknown. So, there is a lot of work ahaed of us.

Areas For Research In Paediatrics

* The standardization of techniques for dose-response curves to indirect stimuli, such as cold air, osmotic challenge, SO2 or metabisulphite and others.
* The development of bronchial responsiveness to direct and indirect stimuli with age in healthy and asthmatic subjects.
* The relative importance of the genetic expression compared to environmental influences on the development of bronchial hyperresponsiveness.
* The significance of the degree of sensitivity to indirect stimuli as a marker of disease activity and predictor of disease outcome. The effect of long-term treatment on direct and indirect stimuli.

If we want to use indirect stimuli as indicators of

bronchial pathology, we need reference values at various ages. It is likely that reference values are different in young children and in adults.

We have to consider the significance of the relative importance of the genetic expression compared to environmental influences on the development of bronchial responsiveness. If we want to do long-term intervention studies, and if we hypothesize that by using medication we can suppress bronchial hyperresponsiveness completely, we must be sure that a dominant genetic factor is not present. If genetic factors are dominant it is far less likely that we will succeed with our medical treatment.

The significance of sensitivity to indirect stimuli and markers of disease activity is not known, and the effect of longterm drug treatment to direct and indirect stimuli has to be evaluated. I realize that research projects in these areas are difficult to carry out because of their long-term character and the fact that often large numbers of subjects are needed, which calls for a multicentre approach.

I want to summarize a few things on reference values at young age, genetic components and effect of long-term drug treatment: Firstly, we speculated on what would happen with bronchial responsiveness between the age of 6 or 7 and 16 years, when airways grow. We calculated that the deposition of methacholine and histamine in the airways will diminish by 30%. So the concentration on the mucosa will diminish, and seemingly bronchial responsiveness (PD20) will become less. We confirmed that in our longitudinal study, and we indeed found that between the age of 7 and 16 bronchial responsiveness drops about 30%. But 30% is within one dose step. So probably this change will not interfere with longitudinal drug studies.

The genetic component is a difficult point. There are studies in favour of a genetic factor, and there are studies against it. I was not able to come to a conclusion reading the literature, and I think this point should be taken up again.

I finally want to comment shortly on drug interference. Our own study is probably the one that lasted for the longest period, i.e., two years. A few other studies are on the way. But to date we do not know anything about the other antiinflammatory drugs like DNCG and nedocromil. We know that these drugs do not modulate bronchial responsiveness. As that may be a question of dose, I think that we need studies with higher doses of these drugs. We know very little about the long-term effect of long-acting beta-2-agonists which may have an antiinflammatory effect also. There are some indications that long-acting beta-2-agonists indeed may suppress bronchial responsiveness to direct stimuli, but these data are scarce. We also have a number of new drugs coming up that need to be tested. That has to be done by standardized methods. It would be good if everybody uses the same kind of technique for that, so that results can be pooled. Finally, nothing is known about airway hyperresponsiveness after discontinuation of long-term treatment. Systemic data are not available, only scattered ones, but nothing systematic.

Discussion

Harnoncourt: I want to ask you a methodical question. You showed us the transcutanuous measurement of oxygen in children while performing provocation tests. Patients after bronchodilatation do not necessarily get an increase in oxygen, they often have a fall. As in your experience FEV1 and transcutanuous oxygen fall were correlated very well, do you think this would be a method for the future?

Kerrebijn: That only works in bronchoconstriction, not in bronchodilatation. That is because of the S-shape of the dissociation curve. So you cannot use this method when you bronchodilate people, but you can use it when you bronchoconstrict them.

Harnoncourt: Do you think we could use It for bronchoprovocation testing?

Kerrebijn: We have done so. We think we can do it, and we now employ this technique in young children in whom we do intervention studies.

Zach: Just along the same line, we have done the same in the last year for cold air challenge correlating conventional pulmonary function tests with oxygen tension measured transcutaneously and found very much the same as was found for histamine and methacholine. A very close relation in children of the fall of PO_2 measured transcutaneously to FEV1.

Kerrebijn: In this respect children are perhaps more easy to investigate than adults because their skin is so thin and therefore transcutaneously PO_2 may reflect better what happens in the blood vessels than in adults.

Bronchial Provocation Testing in Adults - Is there Clinical Relevance?

Anne E. Tattersfield

University of Nottingham, Respiratory Medicine Unit, City Hospital, Nottingham (UK)

In this talk I shall look at the measurement of bronchial responsiveness from the point of view of the patient whose specific concerns are with how their asthma can be treated and what the long term outcome is going to be.

One problem in discussing bronchial reactivity is that it is considered in two ways. Conceptually it is seen as a modol in which hyperresponsiveness represents excessive airway twitchiness with the airway contracting excessively to a given stimulus. In practical terms however we think of bronchial reactivity as a measurement such as the PD20 or PC20, the dose or concentration of histamine or methacholine that causes a 20% fall in FEV1. This is an indirect measurement which we know is influenced by factors other than airway "twitchiness", for example the size of the airways. This is almost certainly one of the reasons why more very young children have a lower PD20 to histamine or methacholine than older children or adults.

What are the potential clinical applications of measurements of bronchial responsiveness? Although it has been suggested that it is useful for the diagnosis of asthma it is very unusual in Britain to measure bronchial responsiveness clinical practice (although it is often measured for research purposes). Are patients in Britain being treated less well because we do not measure bronchial responsiveness regularly? Should we be doing it to make the diagnosis of asthma, or to assess asthma severity or to monitor progress? Would it indicate how the patient is going to progress in the future? Hyperresponsiveness is often said to be one of the hallmarks of asthma and it has been suggested that when patients have symptoms of asthma but no hyperresponsiveness the diagnosis of asthma must be in doubt (1). It has become clear with time, however, that although most patients with asthma have hyperresponsive airways some do not; asthma and hyperresponsiveness (as measured by the response to inhaled histamine or methacholine) are clearly closely associated but do not invariably occur together. I will review the evidence for this view.

The very early studies, such as that by Parker and colleagues in 1965 (2) and subsequent studies by Cockcroft et al (3), and Malo et al (4) showed that when people were studied who were **very** normal and **very** asthmatic (i.e. no doubt about normality or the presence of asthma) you can show complete separation of the two groups in terms of measurements of bronchial responsiveness (PD20 or PC20 FEV1) with no overlap. That of course is a somewhat unreal situation, and one in which you do not need a test because the subjects are clearly asthmatic or clearly normal. If a test is useful, it has to be able to distinguish normal subjects from patients with mild disease, e.g. patients with occasional wheezing or with a non-productive cough and it should be able to separate asthma from diseases such as chronic bronchitis. Looking at extremes does not help to determine whether the test is helpful or not in the relevant clinical situation (5). When bronchial reactivity measurements were made in patients with chronic bronchitis (6), allergic rhinitis without a clear history of asthma (7) or even in cigarette smokers who were asymptomatic (8) or without airflow obstruction (9) the results were found to overlap with the results in patients with asthma and normal subjects. This overlap in measurements of bronchial hyperresponsiveness between different groups means that the sensitivity of the measurement in diagnosing asthma has varied considerably (5), being strongly influenced by the composition of the groups.

One way of overcoming the effect of different composition of the groups is to determine the value of the measurement of bronchial reactivity in a random group of children or adults from the general population, i.e. a cross-section of the whole population. Several studies have done this, particularly in schoolchildren where is it relatively easy to get a good cross-section of children of a particular age and where compliance is high (10-13). One study by Salome and colleagues (11) in 2363 children in Australia showed that 17.9% of all children had bronchial hyperresponsiveness as measured by their histamine challenge test (fig. 1). Over one third of these (6.7%) had never had symptoms of asthma and had never been diagnosed as having asthma. Only 6.7% of the sample had diagnosed asthma, symptoms and bronchial hyperresponsiveness. As can be seen from the figure, there was considerable overlap between symptoms, the diagnosis of asthma and bronchial hyperresponsiveness, but there was a substantial number of children with symptoms or a diagnosis of asthma but no bronchial hyperresponsiveness and some with bronchial hyperresponsiveness who had never had asthmatic symptoms; there was by no means complete agreement. A similar pattern showing overlap but far from complete agreement has been seen in all studies in children (10,12,13).

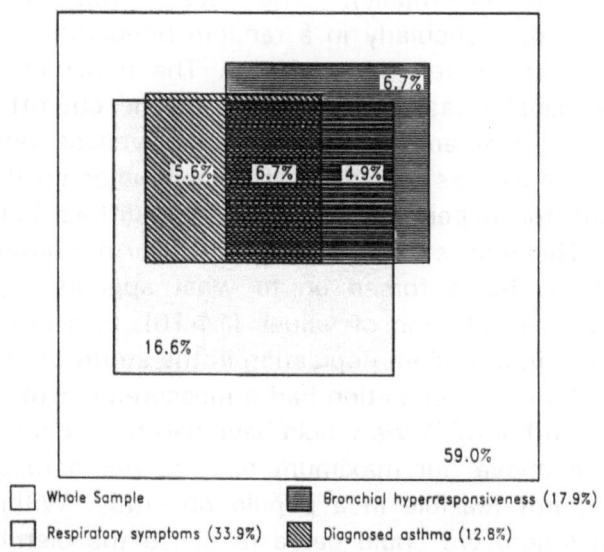

Fig. 1:   Extent of agreement and disagreement between measurement of bronchial responsiveness, symptoms of wheeze and a diagnosis of asthma in **2363** children in Australia. From Salome et al (11).

In a similar study in a random population of adults in

Australia, Woolcock et al (14) showed that the more hyperresponsive you were the more likely you were to have symptoms. But again the agreement was far from complete. Some people with hyperresponsiveness had no symptoms and some with symptoms did not have hyperresponsiveness.

There are problems in measuring bronchial responsiveness which we need to consider. Tests such as histamine and methacholine inhalation are used widely (particularly for adults, many of whom would not be able to exercise vigorously). The dose that can be administered, particularly in a random population, has to be limited to prevent side effects. The maximum dose given is usually assigned arbitrarily as the cut off point between normal and increased responsiveness but this maximum dose has varied in different studies so the cut off point for hyperresponsiveness also differs between studies. This makes comparisons difficult. An arbitrary cut off point is being forced on to what appears to be a continuous distribution of values (14-16). In a study we carried out in a random population in the south of England only 14% of the population had a measurement of PD20, i.e. in the other 86% we would have had to give a dose of histamine above our maximum dose to get a response, which is not feasible in a population study. Within the 14% in whom we could get a response the distribution was unimodal (fig.2), suggesting that we may be looking at the end of a normal distribution of PD20 in the general population though this is not known for sure.

Table 1

	Newcastle children	Dunedin children	Busselton adults	New South Wales children
reference	10	12	14	11
% with wheeze but no BHR	33	35	43	60
% without wheeze but with BHR	33	8	10	12
% BHR with no wheeze/SOB	not available	36	25	37

If you look at the cross-section studies of children in the United Kingdom (Newcastle) (10), Australia (11) and New Zealand (Dunedin) (12) and in adults in New South Wales (14) you can see that the percentage of the normal population that had wheezed and had no bronchial hyperresponsiveness has varied from 33-60% (table 1). On the other hand some adults and schoolchildren had bronchial hyperresponsiveness but no history of wheeze (25-37%). The exact proportion has varied in the different studies because of the different methods used to measure reactivity and the different cut off points used to separate normal from increased responsiveness.

The random population of over 500 adults we studied were aged 18-65 (15). Overall 14% had bronchial hyperresponsiveness but the proportion was slightly greater in the younger and older patients and slightly less in middle age. In the younger subjects atopy (a positive immediate skin test to one or more common allergens) was the most important determinant of bronchial responsiveness; those with atopy being much more likely to be hyperresponsive. In the older age group cigarette smoking was by far the most important factor in determining whether the person was hyperresponsive or not. Atopy was much less common in older subjects and when it was present it was unimportant in determining whether the patient was hyperresponsive or not. The fact that cigarette smoking is associated with hyperresponsiveness is of course one of the problems in using PD20 measurements as a marker for asthma. Several studies have shown that hyperresponsiveness is related to smoking (8,9) and to chronic obstructive airway disease (6). In patients with chronic obstructive airway disease there is a reasonably close relationship between PD20 and FEV1 suggesting that geometric factors related to airway size play a fairly important role in determining hyperresponsiveness in these patients (17). In asthmatics there is a much less close relationship implying that geometric factors are less important and airway "twitchiness" more important in these patients.

The clinical value of reactivity measurements also depends on the assumption that a single measurement on one occasion is representative of measurements on other occasions. In a longitudinal study Joseph et al (18) found that some patients showed a fall in PD20 in relation to increase in symptoms and fall in peak flow rate but others did not. The question then is what action should be taken if symptoms increase and peak flow falls but PD20 remains unchanged. It seems inconceivable that any doctor would ignore the increase in symptoms and fall in peak flow rate, so presumably the fact that PD20 has not changed would be ignored. If PD20 changes with no change in peak flow rate or symptoms is there any justification for altering treatment? None to our knowledge at the moment. Doctors in the United Kingdom would treat the patient's symptoms and airway obstruction and not their bronchial hyperresponsiveness, and this is one reason why they do not see the need to measure bronchial reactivity. It is possible that treating bronchial hyperresponsiveness might improve the long term prognosis, but there is no evidence that this is the case.

What about assessing the severity of asthma? A study by Juniper and colleagues (19) showed a broad relationship between hyperresponsiveness and asthma severity as judged by treatment. But again there was a lot of overlap between groups so the question remains: would you be guided in your treatment by measuring bronchial responsiveness or by the patient's quality of life? Is your treatment going to be guided by a history of waking at night, peak flow recordings and FEV1 or by a measurement of bronchial responsiveness? We are guided by the patient's symptoms and peak flow measurements, since the peak flow meter is measuring actual change in airway caliber whereas bronchial responsiveness is an indirect measurement.

One further question that arises is whether subjects who have no symptoms but have bronchial hyperresponsiveness will develop asthma in the future. A study by Pham and colleagues (20) measured bronchial responsiveness in over 1000 iron mine workers and restudied 800 of them five years later. Subjects who had a positive response to acetylcholine on the first visit were more likely to have a greater fall in FEV1 over the five years than those who did not. If, however, you look at the group who had no symptoms but had a positive acetylcholine test on the initial visit, 91% still had no symptoms and had not developed asthma five years later. Nine percent of this group had developed asthma, so in an epidemiological sense the test had a slight ability to predict the development of asthma. But do you need to know that information? It is only useful if treating bronchial hyperresponsiveness at that stage prevents the development of symptomatic asthma (and outweighs the short term and long term side effects from the treatment). As Professor Kerrebijn has pointed out, there is no evidence yet that that is the case, in either adults or children.

We believe therefore that a measurement of bronchial hyperresponsiveness is of very limited value in adults in making a positive diagnosis of asthma and much less value than regular peak flow recordings. It may be different in children because an exercise test appears to be more specific to asthma than constrictor agents such as histamine. In adults a positive test to histamine tells us that the patient is hyperresponsive to histamine; I do not think it tells us that they have asthma.

I should perhaps emphasize here that measurements of bronchial responsiveness are a very important research tool as they have the great advantage of providing an objective measurement of a function closely related to asthma. Measuring prevalence of asthma is difficult and influenced strongly by local diagnostic practices; even measuring the prevalence of symptoms such as wheeze may be influenced by cultural factors or problems with translation when trying to compare the prevalence of asthma, for example, in different countries. I will quote a recent example of the value of bronchial reactivity tests from a study of asthma prevalence in over 2000 children in Zimbabwe. The children came from Northern Harare, Southern Harare (which is much poorer) and a rural area. The children carried out an exercise test (running around a field as fast as they could for six minutes). The number that developed exercise-induced bronchoconstriction was very different in the three areas, 5.8% in Northern Harare, 3.1% in Southern Harare and 0.1% in the rural area. This objective information provides much more convincing evidence of the importance of environmental factors in determining asthma prevalence (since all the children are from the same genetic background) than a questionnaire measurement of diagnosed asthma or even symptoms because it is not subject to bias due to language problems or diagnostic habits in Zimbabwe. Bronchial reactivity measurements provide an objective measure for studies trying to determine which environmental factors are important for the development of asthma.

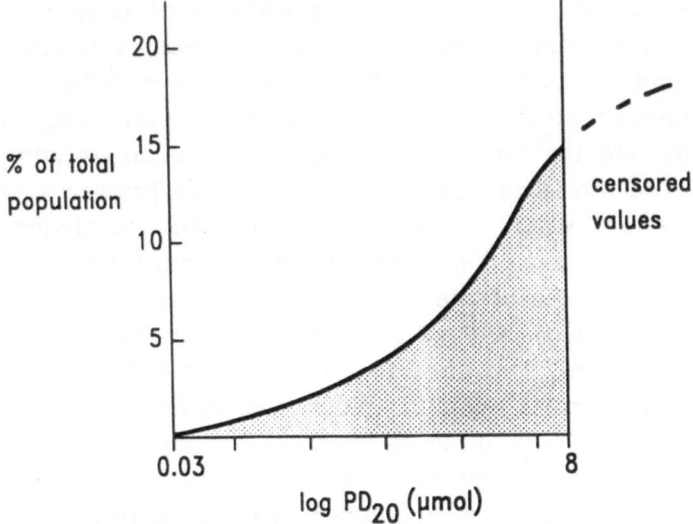

Fig. 2: Distribution of bronchial responsiveness in a community based cross-section of adults in Hampshire and Dorset, England. The highest dose given was 4 umol (extrapolated to 8 umol). 86% of the population did not bronchoconstrict to 8 umol (described as censored values). From Burney et al (15).

In our study in Hampshire and Dorset we were able to pursue the hypothesis suggested by Burney (22) that salt intake may be important in determining the prevalence of asthma (fig.2). We measured urinary sodium output in 153 men and found quite a large difference in bronchial responsiveness in relation to sodium output (23). I doubt whether we would have found a relationship if we had measured asthma according to symptoms or a known diagnosis of asthma and the findings would have been less convincing. Thus measuring bronchial responsiveness has important research uses, but apart from certain limited situations I am not convinced that it has a useful role in clinical practice.

Discussion

Zwick: Can you tell us how often the provocation test is negative in symptomatic asthmatics, - the methacholine challenge test for example?

Tattersfield: In symptomatic asthmatics?

Zwick: Yes. You showed considerable overlap but nobody knows exactly how much.

Tattersfield: The figures differ in different studies, partly because people have used different criteria for symptomatic asthma and partly because they are using different tests and different cut-off points. It varies in the different studies as shown in table 1 but in population studies is as high as 60%. It depends very much on what you mean by symptomatic asthma; in some of the population studies the subjects may be having symptoms very infrequently.

Zwick: I am asking for that person that tells you having for example dyspnoea at nights. In your office the ventilation parameters are in the normal range, so you doubt if he is asthmatic or not. In that case I would do an inhalation challenge to get a quick answer, thinking that the number of patients having asthma and a negative bronchial provocation test is very low. I thought there are about one or two percent of patients with a negative methacholine inhalation challenge test having asthma.

Tattersfield: You may be right if you are talking about somebody who is getting symptoms at night. There is no hard information on that particular group of patients as far as I am aware.

Zwick: I think that is what we are asked very often. That is why we are doing inhalation challenge tests in clinical practice. To diagnose a symptomatic asthmatic is not very difficult but it is difficult to exclude it. I thought that the exclusion criterium is good. So in spite of the fact that methacholine challenge shows a negative result in my opinion there is only one or two percent possibility that this person has asthma.

Tattersfield: Yes, but there is no gold standard to test whether you are right or not. You have to be careful not to be in a self-fulfilling situation. If you say that nearly all subjects with a positive test have asthma, then your statement has to be right because your are making the diagnosis on the basis of the test.

Kerrebijn: The point is again: what does it mean that you have a positive or a negative test? In my conception, a patient with symptoms and a negative methacholine or histamine challenge test has either inflammation at a low level or epithelial damage or probably no smooth muscle hypertrophy or no changes in elastic load. All these abnormalities which determine the effects of methacholine or histamine challenge are so limited that we cannot detect them with the rather rough tests which we apply. So I would not be too much concerned about these patients, and I would treat them symptomatically.

What does it mean that the patient has no symptoms but a positive test? That may mean - but we do not know that exactly - that for instance acute inflammation is at a very low level. But it also may well be that other factors like smooth muscle hypertrophy is present or elastic load or thickness of the airways have already permanently changed. There are few data now existing from studies that are going on, and which have been presented at the latest ATS-meeting. I can imagine that a negative test does not mean that there is no pathology if there are symptoms. But all that is not really known.

Tattersfield: People are now looking at bronchial biopsies and bronchial alveolar lavage in relation to bronchial responsiveness and I understand also that there is a relationship between the two although agreement is not particularly close. Some people with quite marked hyperresponsiveness apparently do not have much inflammation in the airways that were biopsied (which may not necessarily be representative). We need more information on that.

Hargreave: I would like to say something about the same point. If the person has current symptoms and the test is negative, it seems very unlikely that there is any variable airflow obstruction. Because if you measure diurnal variability of peak flow rates in these people, they are usually normal. So what it means clinically is that that particularly person does not need treatment with a bronchodilator. It does not exclude the fact that the symptoms could be associated with airway inflammation and an increase in eosinophils and would respond to steroids. But it means that if the test is negative and there are symptoms, one has to think about other causes for the symptoms, and there are other causes like hyperventilation or bronchitis etc. There is a place for the test in diagnosis, and it depends on what sort of patients you are seeing. The majority of asthmatics are mild. The majority therefore represents with symptoms having a normal spirometry and you want to either exclude or support the diagnosis of asthma. The test is an easy test to do, it is done on the supervision, in a controlled situation. I think it is useful to give you an idea whether the person has or has not variable airflow obstruction in association with symptoms.

Kerrebijn: At what level of airway inflammation will methacholine or histamine tests become positive?

Hargreave: It does not relate necessarily to the degree of inflammation and we do not know why but that is based on what Dr. Ingram was showing and what has been shown in pathological studies in chronic airflow limitation. It is likely that the levels of inflammation in those patients are different. Perhaps it is involving the larger airways and not involving the smaller ones. If one has not got involvement of the smaller airways, one is more likely to have normal airway function and normal responsiveness. I think that is probably what is happening, but it needs establishing.

Kerrebijn: In this respect probably the maximal plateau level is as important as the shift, because that may indicate whether there are permanent changes, i.e., structural changes in the airway wall.

Hargreave: In a group of subjects we have done some further studies, where we have been able to measure the level of airway responsiveness. When you can do that and you treat the person with inhaled steroid, the level of responsiveness improves to the normal range.

Tattersfield: The patients that are referred to my clinic usually have a low FEV1 and they would usually have bronchial hyperresponsiveness. When patients have a normal FEV1 we give them a peak flow meter and ask them to measure peak flow rate at home. That seems to me to be a more direct measurement and more useful than the indirect measurement of bronchial provocation testing. Studies that have compared peak flow variability with bronchial responsiveness have shown a relationship but not a very close one; the correlation coefficient is usually at about 0.5, even in the study from Hamilton (Ryan et al, Thorax 1982; 37:423-429).

Hargreave: No, it is higher than that if you are measuring diurnal variation of peak flow rates. In order to get a good relationship you have to include the response to a bronchodilator. It is a measurement which is made from the lowest value (usually in the morning) before bronchodilator and the highest value after bronchodilator (usually at the end of the afternoon), expressed in relation to the highest value. When you do compare that with bronchial responsiveness done by inhalation studies, you get a very nice correlation between the PC20 and a degree of variability of flow rates.

Tattersfield: What was your correlation coefficient then?

Hargreave: It was about 0.8; In fact if you take these people who are asymptomatic and say they had never had symptoms of asthma in the past and you demonstrate that they are hyperresponsive they usually show hyperresponsiveness within a mildly increased range as you showed it from Ann Woolcock's study. If you do look at diurnal variability of peak flow rate in those - it is the same in the asymptomatic subjects as it is in the symptomatic subjects matched for PC20 - you will find a difference to what you see in the normal subjects. So I think the peak flow variability and the PC20 are really telling you virtually the same thing.

Tattersfield: I think again one of the problems is the question of selection of patients, as in the early Parker study. In our community based survey (Higgings et al, Am Rev Respir Dis 1989; 140:1368-1372) there was quite a lot of overlap of peak flow variability between subjects who gave a very clear history of asthma and normal people that had never had any symptoms. We did not include the response to a bronchodilator, but we recorded peak flow values as they are normally recorded.

Popp: Some authors in Australia believe that they can assess the severity of asthma or if asthma is life threatening by the degree of bronchial hyper-responsiveness. Do you believe it?

Tattersfield: I suspect that the people with more bronchial hyperresponsiveness are more at risk from life threatening asthma. Patients with brittle asthma are at greater risk of life threatening asthma, and I suspect that on the whole they have greater bronchial hyperresponsiveness than patients with non-brittle asthma. We have completed a study to look at this and should have some answers soon.

Kummer: I think this statement of yours finds support in a recent paper by Gilbert and and Auchincloss (Chest 1990, 97:562). They had taken the data of Casale et al (J Appl Phys 1987, 62:1888, and J Appl Phys 1988, 64:2258) and compared clinical probability with the result of a methacholine provocation test in non-atopic non-smoking and smoking asthmatics with a large scope of different PD20 levels. They find the lowest-dose responders as having an over 90% post-test probability of asthma, regardless of what the clinical impression was. On the other side, asthmatics with a very high clinical suspicion on the first place, any positive BPT is ranked as high post-test probability, regardless of how high the test-dose was. For all measurements between these extremes they calculate sensitivities and specificities, which might be a practicable aid in decision finding for a given patient, provided you know enough about the history and concomitant clinical findings, including smoking habits.

Tattersfield: How does he know that he was right? Again this depends on how asthma is defined. Where is the gold standard for saying it is asthma?

Kummer: Instead of the gold standard - which does not exist - they apply Bayes' theorem to calculate the post-test probability of asthma, based on pre-test probability (clinical impression) and the test result (threshold dose of methacholine). From the diagrams, which they had based not only on Casale's data, they stress the importance of the impact of clinical data on the diagnosis of asthma, and they express their caveat for "depersonalized" evaluation of the test results.

Madjar: What is the percentage of real asthmatics having a negative bronchial provocation test in a symptom free period without any treatment?

Tattersfield: These questions are a bit like asking how long is a piece of string. I do not know the answer. The longer a patient has been free of symptoms, the higher the PD20 is likely to be.

Madjar: If you have a patient with unknown diagnosis and there is no hyperresponsiveness estimated by methacholine, I think we can exclude asthma with a high probability. In my impression it is not more than 5 to 6 percent of asthmatics that are non-reactive without any treatment. I also do not know this exactly, but I can agree with a former speaker who told that it could be 1 or 2 percent, not more.

Tattersfield: Well, that may be true for the population you are looking for. I am just trying to make the point that it depends on the population you investigate.

Madjar: On the other side of course it is not allowed for us to say that somebody is a real asthmatic if he has a positive reaction to methacholine. Perhaps not more than 30% of those who are hyperresponsive on a low dosage are asthmatics. Several other tests, for example with distilled water, hypertonic sodiumchloride or potassium chloride have a better specificity for asthma. We use methacholine tests for diagnosing asthma because these tests are more sensitive.

Tattersfield: I think that is a good point. The question is whether a test such as hypertonic saline or destilled water might be more specific for asthma and that may be true. But at the moment these tests are still at the stage of the Parkers studies for histamine. In other words: people have still tended to look at them in selected populations. I am not aware of any studies that have looked at these tests in a community population apart from cold air hyperventilation. When cold air was used in a community population there was overlap (Weiss et al, Am Rev Respir Dis 1984; 129:898-902).

Wieser: I am very interested in the papers about salt intake. Can you tell us a little bit more about this topic, can you improve symptoms of asthma if you reduce salt intake?

Tattersfield: The two studies that have been published show that you can reduce bronchial responsiveness by about one doubling dose of histamine, if you go onto a low salt diet. (Burney et al, Thorax 1989; 44:36-41; Javaid et al, Br Med J 1988; 297:454). The reduction in salt intake was large, and I do not know how many patients would stick to it. It is very interesting from the point of view of mechanisms and would explain some of the epidemiological data about asthma which seems to be more common in Western society.

References

1. Adelroth E, Hargreave FE, Ramsdale EH: Do physicians need objective measurements to diagnose asthma? Am Rev Respir Dis (1986) 134:703-707
2. Parker CD,Bilbo RE, Reed CE: Methacholine aerosol as test for bronchial asthma. Arch Intern Med (1965) 115:452-458
3. Cockcroft DW, Killian DN, Mellon JJA, Hargreave FE: Bronchial reactivity to inhaled histamine: a method and clinical survey. Clin Allergy (1977) 7:235-243
4. Malo J-L, Pineau L, Cartier A, Martin RR: Reference values of the provocative concentrations of

methacholine that cause 6 & and 20% changes in forced expiratory volume in one second in a normal population. Am Rev Respir Dis (1983) 128:8-11

5. Britton JR, Tattersfield AE: Does measurement of bronchial hyperreactivity help in the clinical diagnosis of asthma? Eur J Respir Dis (1986) 68:233-238

6. Ramsdell JW, Nachtwey FJ, Moser KM: Bronchial hyperreactivity in chronic obstructive bronchitis. Am Rev Respir Dis (1982) 126:829-832

7. Stevens WJ, Vermiere PA: Bronchial responsiveness to histamine and allergen in patients with asthma, rhinitis, cough. Eur J Respir Dis (1980) 61:203-212

8. Cockcroft DW, Berscheid BA, Murdock KY: Bronchial response to inhaled histamine in asymptomatic young smokers. Eur J Respir Dis (1983) 64:207-211

9. Gerrard JW, Cockcroft DW, Mink JT, Cotton DJ, Poonawala R, Dosman JA: Increased non-specific bronchial reactivity in cigarette smokers with normal lung function. Am Rev Respir Dis (1980) 122:577-581

10. Lee DA, Winslow NR, Speight ANP, Hey EN: Prevalence and spectrum of asthma in childhood. Br Med J (1983) 286:1256-1258

11. Salome CM, Peat JK, Britton WJ, Woolcock AJ: Bronchial hyperresponsiveness in two populations of Australian schoolchildren. Clin Allergy (1987) 17:271-281

12. Sears MR, Jones DT, Holdaway MD, Hewitt CJ, Flannery EM, Herbison GP, Silva PA: Prevalence of bronchial reactivity to inhaled methacholine in New Zealand children. Thorax (1986) 41:283-289

13. Pattemore PK, Asher MI, Harrison AC, Mitchell EA, Rea HH, Stewart AW: The interrelationship among bronchial hyperresponsiveness, the diagnosis of asthma, and asthma symptoms. Am Rev Respir Dis (1990) 142:549-554

14. Woolcock AJ, Peak JK, Salome CM, Yan K, Anderson SD, Schoeffel RE, McGowage G, Killalea T: Prevalence of bronchial hyperresponsiveness and asthma in a rural adult population. Thorax (1987) 42:361-368

15. Burney PGJ, Britton JR, Chinn S, Tattersfield AE, Papacosta AO, Kelson MC, Anderson F,Corfield DR: Descriptive epidemiology of bronchial reactivity in an adult population: results from a community study. Thorax (1987) 42:361-368

16. Cockcroft DW, Berscheid BA, Murdock KY: Unimodal distribution of bronchial responsiveness in a random human population. Chest (1983) 83:751-754

17. Pride NB, Taylor RG, Lim TK, Joyce H, Watson A: Bronchial hyperresponsiveness as a risk factor for progressive airflow obstruction in smokers. Clin Respir Physiol (1987) 23:369-375

18. Josephs LK, Gregg I, Mullee MA, Holgate ST: Nonspecific bronchial reactivity and its relationship to the clinical expression of asthma. Am Rev Respir Dis (1989) 140:350-357

19. Juniper EF, Frith PA, Hargreave FE: Airway responsiveness to histamine and methacholine: relationship to minimum treatment to control symptoms of asthma. Thorax (1981) 36:575-579

20. Pham QT, Mur JM, Chau N, Gabiano M, Henquel JC,

Teculescu D: Prognostic value of acetylcholine challenge test: a prospective study. Br J Ind Med (1984) 41:267-271

21. Keeley DJ, Neill P, Gallivan S: Comparison of the prevalence of reversible airways obstruction in rural and urban Zimbabwean children. Thorax (1991) 46:549-553

22. Burney PGJ: A diet rich in sodium may potentiate asthma: epidemiological evidence for a new hypothesis. In: Holland WW, ed. Proceedings of Fogarty International Center workshop on etiology of asthma, National Institutes of Health, 25-27 June 1985

23. Burney PGJ, Britton JR, Chinn S, Tattersfield AE, Platt HS, Papacosta AO, Kelson MC: Response to inhaled histamine and 24 hour sodium excretion. Br Med J (1986) 292:1483-1486

Principles and Approaches in Asthma Therapy

Friedrich Kummer

Wilhelminenspital der Stadt Wien, Vienna (Austria)

The preceding presentations have dealt extensively with the interconnections of hyperresponsiveness and asthma. Before going into details of asthma therapy, let me outline the terms and meaning of **pathogenesis** and **pathophysiology.**

Pathogenesis stands for the transition of a system totally normal both in function and structure, into any abnormal state.

Pathophysiology, in turn, is everything that follows, based on pathogenesis. In case of asthma, this is the first assessable change of function (hyperresponsiveness) and/or structure (e.g., eosinophilia). Even from here it is some way to go to clinical manifestations in symptoms and signs.

The most elegant **therapeutic** approach would be a successfull reversal of pathogenesis, which would be the one and only causal therapy. A semi-causal therapy is still a useful thing, but only of limited potential quo ad "cure", because it modifies pathophysiology, but leaves pathogenesis untouched.

On the contrary, **clinical** management is principally **symptomatic** treatment. It should be stressed at this point - as it was demonstrated in some previous papers - that medicamentous downregulation of clinical symptoms has an impact on pathophysiology (e.g., inhaled steroids and inflammation).

Based on these considerations, the development of bronchial hyperresponsiveness can be looked at pathogenetically as well as pathophysiologically. I do not believe that one can talk about pathogenesis of asthma without talking about **pathogenesis of bronchial hyperresponsiveness.** Does it develop from viral infection in childhood? Are children born with a disposition for infection? Is there an inborn "narrow bronchial system", i.e., a different geometry of growing airways in "wheezy" children? It seems to be the old question of the chicken and the egg: is the narrow bronchus guilty of hyperreactivity, or vice versa?

Atopy has for a long time been dealt with as the major pathogenesis of hyperresponsiveness. But then we learned about **genetic factors,** which are the even deeper basis of atopy and allergy on the one side, and hyperresponsiveness on the other side. Then we are taught that oil particles and other **pollutants** can activate rather innocent pollen to become an allergen, and that - on the other hand - diesel exhaust particles can cause "nonspecific" increases in IgE-production.

Hyperresponsiveness per se is not a disease, it rather belongs to phenomena which carry some predisposition for asthma. Prof. Tattersfield mentioned six hyperresponsive individuals, who did not develop a trace of asthma in a longitudinal study over five years. So it seems, that the state of hyperresponsiveness still needs a trigger, which eventually turns latent (subclinical) hyperresponsiveness into clinically manifest asthma.

Certainly, the degree of hyperresponsiveness and the method, by which it was assessed, play a role in this game, as well as the mode and dose of the trigger itself. Short, high-dosed allergen inhalation does rarely cause sensitization, unless we have an extremely sensible atopic individual.

The first aim of therapy logically should be the attenuation of hyperresponsiveness. Effective medications have been developed to interfere with synthesis, liberation and target-organ-effect of those mediators, which cause and perpetuate hyperresponsiveness. Some medications are capable of all this, some only in part. One can also focus on the triggers which are relevant in the environment of an individual asthmatic in order to ban them more or less specifically. And we are able to up-regulate the beta-2-receptor, so nothing can be of harm for it, whatever happens in the mucosa in terms of mediator release or direct irritation.

The **anti-mediator-approach**: It came quite clear, that steroids in general, and the inhaled ones in particular, can interfere with synthesis, as well as liberation and potency of mediators, hence being capable of downregulating hyperresponsiveness. Prof. Kerrebijn has told us, however, that it can not be totally and lastingly abolished: The curve of threshold goes up promisingly, but it does not become normal, and the open question remains, how long it will stay that way after therapy has been stopped. Many laboratories around the globe engage in investigating specific mediator blockers. Prof. Barnes has named some of the **anti-leukotrienes**, which are acting at the central site of airway inflammation and perpetuating the subclinical basis of asthma. However, clinical applicability has not yet ensued.

We have already collected positive experience with **cromone-derivatives**, which seem to block mediator release in a more non-specific way, whereby hyperresponsiveness is somewhat modulated, but certainly not abolished. DNCG is controversal in diminishing hyperresponsiveness, our hopes now focus on nedocromil, which seems to be more potent, especially also in non-atopic individuals. Ketotifen seems to have a temporary and rather short term effect and seems to block PAF in particular.

A more recent concept is the **longacting beta-2-stimulation**, which seems to act beyond the bronchodilating, merely symptomatic therapy. We have reason to hope that, e.g., salmeterol might have some direct or indirect antiinflammatory action. At present, however, we are left with a promising theory which carries too little evidence as to suggest a substitution for steroids and their direct antiinflammatory action.

Up to this point we have mentioned methods and ways to influence hyperresponsiveness and broncho-constriction. But there is a good deal left over to do for the treating physician to identify and cut out **the triggers**, which are responsible for individual attacks or for the smoldering inflammation itself. The most prominent trigger is undoubtedly the **allergen**. We know how important it is to avoid exposure by all means, which can reasonably be applied. Sometimes we have to find a compromise between a radical and a more permissive attitude to accomplish this, a difficult task which can be solved only by a sound personal communication between doctor and patient.

Infection, especially by viral organisms, can enhance hyperresponsiveness and symptoms. But how can we safely practice attenuation of proneness for infection, to an extent that we can appreciably cut down on asthmatic attacks? There are some promising attempts with non-specific immunemodulation (upregulation of IgA, downregulation of IgE), but the bridge to significant changes in the number of attacks has not been built so far. Similar considerations apply to antibiotics, which are of limited indication when we assume that viral infections are much more dangerous for asthma than bacterial ones.

In spite of **physical exercise** being proven as a trigger, it would not make sense to avoid it. On the contrary, patients with exercise induced asthma should be advised to move around, to do sports of the endurance type, and to use proper premedications before engaging in strenuous exercise. Avoidance of sprint-like exercise and warming of inspired air serve as additional tricks.

Chemicals are of predominant importance, when they occur in professional or hobby work environments, but their role as air pollutants is controversial, as to what the triggering of asthmatic attacks is concerned. In isocyanate-exposed workers with asthma, it seems that most of them have acquired lifelong non-specific hyperreactivity, as it was documented by Prof. Fabri in industrial workers many years after work cessation.

The role of **tobacco smoking** in establishing or maintaining hyperresponsiveness is not quite clear: there is a high coincidence of hyperresponsiveness and smoking, and there seems to exist an effect of environmental smoke on asthmatic symptoms, which shows good reproducibility and is clearly non-allergic. When hyperresponsive smokers quit smoking, they lack a lasting change of their hyperresponsive threshold.

Another problem is the impact of **gastro-esophageal reflux** during the night. It seems, that a small group of asthmatics with proven reflux-based complaints can have benefit of consequent treatment, particularly if symptoms are confined to the night time. These patients may be different from the "morning dippers", who experience their symptoms usually after four o'clock, while the reflux-asthma manifests at any time, sometimes immediately after going to bed. The more symptoms during the day, the less the chances of cure even by "total" reflux management (antacids, H2-blockers, tilting the bed, avoidance of late meal, not to speak of surgical correction).

Psychological triggers are usually based on conditioning effects by smells and hyperventilation. The role of personality is more than controversial. The treatment of psychological triggers is a wide playground for psychiatrists and alternative methods, but cures are extremely rare. The best psychic treatment of an asthmatic consists of the proper training of optimal selfmedication, by which the reliability on the medication can be experienced by the patient, thus creating more emotional stability, self esteem, and reduction of panic reactions when the doctor is out of reach.

Finally we have to mention the **medicamentous trigger**, which is often iatrogenic, but even more often caused by uncontrolled self-medication with analgetics of the aspirin or NSAID-type. The iatrogenic triggers in this context consist of betablockers and salicylates, in cardiocirculatory indications as well as in treatment of glaucoma (Timolol). It should be stressed that the small doses applied (baby-aspirin respectively eye-drops) rarely cause severe attacks, but rather worsen lung function, which is hardly perceived as such, but causes considerable shortness of breath on exercise. Sometimes these symptoms come to the knowledge of the patient only in retrospect after they were diagnosed as to their cause and relieved by change of therapy.

In summary, the only causal treatment of asthma should aim at the extinction of inflammation and hyperresponsiveness. This is, at present, utopia, but with our limited tools we can diminish the pathophysiological sequelae by prevention of attacks (semicausal) and dilating the constricted airways. A special point in long-term management of asthma is identification and treatment of individual triggering mechanisms of asthma, despite their great variety in strength and relevance. Comprehensive asthma treatment encompasses measures to diminish hyperresponsiveness (steroids, mediator-blockade), long acting bronchodilation, and "trigger-ectomy", which is the fruit of individual and intensive talks and information exchanged by the patient and his/her physician.